McGraw-Hill

My Math

Grade 1

Think Smart for the Smarter Balanced Assessment

www.mheonline.com

McGraw-Hill is committed to providing Instruction materials in Science, Technology, Engineering, and Mathematics (STEM) that give all students a solid foundation, one that prepares them for college and careers in the 21st century.

Send all inquiries to:
McGraw-Hill Education
8787 Orion Place
Columbus, OH 43240

ISBN: 978-0-02-131179-8
MHID: 0-02-131179-X

Printed in the United States of America.

2 3 4 5 6 7 8 9 10 RHR 20 19 18 17 16 15

Our mission is to provide educational resources that enable students to become the problem solvers of the 21st century and inspire them to explore careers within Science, Technology, Engineering, and Mathematics (STEM) related fields

Contents

Teacher's Guide to Smarter Balanced Assessment Preparation

Whether it is the print ***Think Smart for Smarter Balanced Assessment*** or online at **ConnectED**.mcgraw-hill.com, *McGraw-Hill My Math* helps students prepare for the SBAC testing.

How to Use this Book

Think Smart for the Smarter Balanced Assessment includes experiences needed to prepare students for the upcoming SBAC assessment. The exercises in this book give students a taste of the different types of questions that may appear on the SBAC test.

Smarter Balanced Assessment Item Types

- Familiarizes students with commonly-seen item types
- Each type comes with a description of the online experience, helpful, hints, and a problem for students to try on their own.

Countdown to SBAC

- Prepares students in the 20 weeks leading up to the SBAC assessment
- Consists of five problems per week, paced with order of the *McGraw-Hill My Math Student Edition* with built in review.
- **Ideas for Use** Begin use in October for pacing up to the beginning of March. Assign each weekly countdown as in-class work for small groups, homework, a practice assessment, or a weekly quiz. You may assign one problem per day or have students complete all five problems at once.

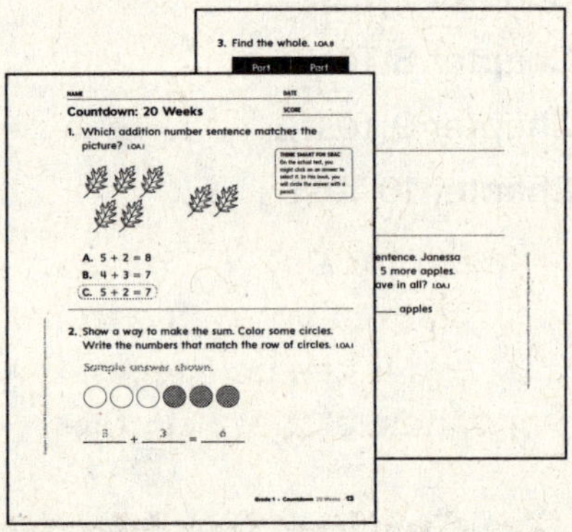

Chapter Tests

- Each six-page test assesses all of the Common Core State Standards for Mathematics presented in the chapter.
- Each question mirrors an item type that might be found on the SBAC assessment, including multi-part questions.
- **Ideas for Use** Assign as in-class group work, homework, a practice assessment, a diagnostic assessment before beginning the chapter, or a summative assessment upon completing the chapter.

Chapter Performance Tasks

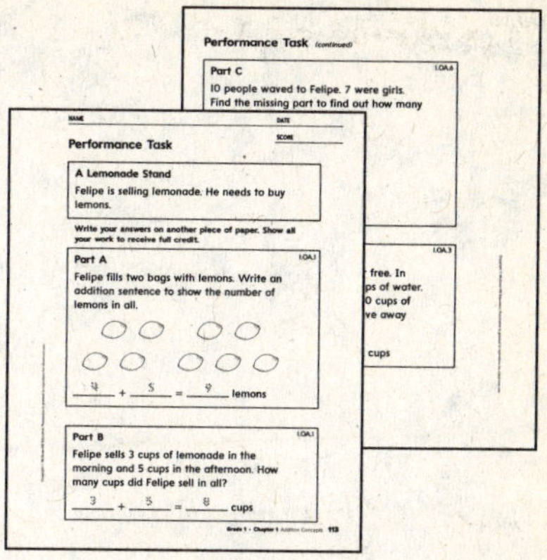

- Each two-page performance task measures students' abilities to integrate knowledge and skills across multiple standards. This helps students prepare for the rigor expected in college and future careers.
- A rubric describes the standards assessed and guidelines for scoring student work for full and partial credit.
- Sample student work is also included in the answer section of this book.
- **Ideas for Use** Assign as in-class small group work, homework, a practice assessment, or in conjunction with the Chapter Test as part of the summative assessment upon completion of the chapter.

Benchmark Tests

Four benchmark tests are included in this book. All problems on the tests mirror the item types that may be found on the SBAC assessment. Each benchmark test also includes a performance task.

- The ***first*** benchmark test is an eight-page assessment that addresses the CCSS from the first third of the Student Edition.
- The ***second*** benchmark test addresses the second third of the Student Edition.
- The ***third*** and ***fourth*** benchmark tests (Forms A and B) are twelve-page assessments that address the CCSS from the entire year, all chapters of the Student Edition.
- A rubric is provided in the Answer section for scoring the performance task portion of each test.
- **Ideas for Use** Each benchmark test can be used as a diagnostic assessment prior to instruction or as a summative assessment upon completion of instruction. Forms A and B can be used as a pretest at the beginning of the year and then as a posttest at the end of the year to measure mastery progress.

Go Online for More! connectED.mcgraw-hill.com

Additional question analysis for each item in this book is available online. The analysis includes SBAC claims, DOK levels, CCSS assessed, Mathematical Practices implemented, and more.

Performance Task rubrics to help students guide their responses are also available. These describe the tasks students should perform correctly in order to receive maximum credit.

Additional year-end performance tasks are available for Grades 1 through 5 in ***Countdown to Common Core*** blackline masters available under Assessment in ConnectED.

Students can also be assigned tech-enhanced questions from the eAssessment Suite in ConnectED. These questions provide not only rigor, but the functionality students may experience when taking the online SBAC assessment.

Smarter Balanced Assessment Item Types

In Grade 3, students will begin to take the Smarter Balanced Test for Math. It is a test that is taken on a computer. The problems on the next few pages show the kinds of questions students might have to answer and what to do to show the answer on the computer. However, the student materials in the Grade 1 book center mainly on ***Multiple Choice, Drag and Drop***, and ***Click to Select*** formats.

Selected Response means that you are given answers from which you can choose.

Selected Response Items

Regular multiple choice questions are like tests you may have taken before. Read the question and then choose the <u>one</u> best answer.

Multiple Choice

Sasha has a collection of coins. There are 3 in one pile and 2 in another. How many coins does she have in all?

- ☐ 2
- ☐ 3
- ☐ 5

ONLINE EXPERIENCE Click on the box to select the one correct answer.

HELPFUL HINT Only one answer is correct. You may be able to rule out some of the answer choices because they are unreasonable.

Sometimes a multiple choice question may have more than one answer that is correct. The question may or may not tell you how many to choose.

Multiple Correct Answer

A fact family is made from the numbers 3, 7, and 10. Choose **two** statements that belong to this fact family.

- ☐ $3 + 7 = 10$
- ☐ $10 - 3 = 7$
- ☐ $5 + 5 = 10$
- ☐ $10 - 6 = 4$

ONLINE EXPERIENCE Click on the box to select it.

HELPFUL HINT Read each answer choice carefully. There may be more than one right answer.

Another type of question asks you to tell whether the sentence given is true or false. It may also ask you whether you agree with the statement, or if it is true. Then you select yes or no to tell whether you agree.

Multiple True/False or Multiple Yes/No

There were no birds on the pond. Two brown ducks swim into the pond. Five geese fly in and land on the pond. Tell whether you agree with each sentence.

ONLINE EXPERIENCE Click on the box to select it.

HELPFUL HINT There is more than one statement. Any or all of them may be correct.

Yes	No	
☐	☐	There are more geese than ducks on the pond.
☐	☐	There are 2 birds on the pond.
☐	☐	5 geese flew away.
☐	☐	There are 7 birds in all on the pond.

You may have to choose your answer from a group of objects.

Click to Select

Which model shows $4 - 1 = 3$?

ONLINE EXPERIENCE Click on the figure to select it.

HELPFUL HINT On this page you can draw a circle or a box around the figure you want to choose.

When no choices are given from which you can choose, you must create the correct answer. One way is to type in the correct answer. Another may be to make the correct answer from parts that are given to you.

Constructed-Response Items

Fill in the Blank

What time is shown on the clock?

Write the time on the digital clock.

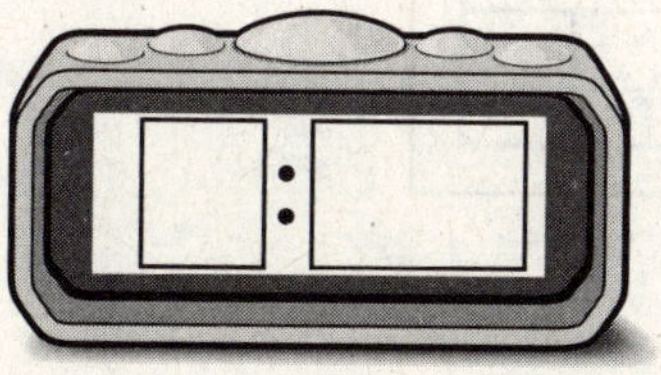

ONLINE EXPERIENCE You will click on the space and a keyboard will appear for you to use to write the numbers and symbols you need.

HELPFUL HINT You must write all of the digits for the time. For example, do not forget to include the minutes if the time shown is "6 o'clock."

Sometimes you must use your mouse to click on an object and drag it to the correct place to create your answer.

Drag and Drop

Two days were sunny.

Six days were rainy.

Five days were cloudy.

Use the pictures to make a picture graph.

ONLINE EXPERIENCE You will click on an object and drag it to the spot it belongs.

HELPFUL HINT You can use S, R, or C to stand for each type of day.

Daily Weather

Sunny						
Rainy						
Cloudy						

NAME DATE

Countdown: 20 Weeks

SCORE

1. Which addition number sentence matches the picture? I.OA.I

THINK SMART FOR SBAC
On the actual test, you might click on an answer to select it. In this book, you will circle the answer with a pencil.

A. 5 + 2 = 8

B. 4 + 3 = 7

C. 5 + 2 = 7

2. Show a way to make the sum. Color some circles. Write the numbers that match the row of circles. I.OA.I

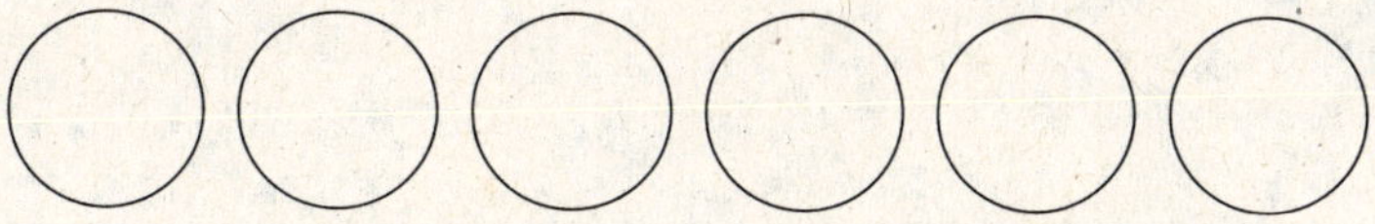

_______ + _______ = _______

3. Find the whole. 1.OA.8

Part	Part
1	4
Whole	

4. Add. 1.OA.6

$$\begin{array}{r} 2 \\ +\ 2 \\ \hline \end{array}$$

5. Write an addition number sentence. Janessa picks 4 apples. Armani picks 5 more apples. How many apples do they have in all? 1.OA.1

________ + ________ = ________ apples

NAME ____________________ DATE ____________

Countdown: 19 Weeks

SCORE ____________

1. How many in all? 1.OA.1

A. 5

B. 6

C. 7

2. Find the missing part of 10. 1.OA.8

Part	Part
2	
Whole	
10	

THINK SMART FOR SBAC
On the actual test, you might be asked to use a keyboard to enter the number in the box. In this book, you will write the number with a pencil.

3. Add. 1.OA.6

2 + 5 = ________

4. True or false? 1.OA.1, 1.OA.7

Rolando finds 2 rocks at the beach. He finds 3 rocks in his yard. Rolando has 5 rocks.

True　　　False

5. Write an addition sentence. 1.OA.1

______ + ______ = ______

NAME DATE

SCORE

Countdown: 18 Weeks

1. A juggler is juggling 5 balls. He drops 2 balls. How many are left? 1.OA.1

_____ balls

2. Subtract to fill in the missing part. 1.OA.1

Part	Part
1	
Whole	
7	

THINK SMART FOR SBAC
On the actual test, you may be asked to click into the table and type your answer. In this book, you will write your answer in the table using a pencil instead.

3. There are 8 bugs crawling. 4 bugs stop. Write a subtraction number sentence to find the number of bugs still crawling. 1.OA.1

_______ ◯ _______ = _______ bugs

4. Subtract. 1.OA.3

$9 - 9 =$ _______

5. There are 6 bunnies. 2 bunnies hop away. Cross out the bunnies to find how many bunnies are left. 1.OA.6

$$\begin{array}{r} 6 \\ -\ 2 \\ \hline \end{array}$$

NAME DATE

Countdown: 17 Weeks

SCORE

1. Write a subtraction sentence. 1.OA.6

_______ – _______ = _______

2. Add. 1.OA.1

5 + 1 = _______

THINK SMART FOR SBAC
On the actual test, you may be asked to click on the blanks and type the numbers. In this book, you will write the numbers using a pencil instead.

3. There are 6 birds in the tree. There are 2 birds on the ground. Write a subtraction number sentence to find how many more birds there are in the tree than on the ground. I.OA.I

________ ◯ ________ = ________ birds

4. Subtract. I.OA.6

5 − 2 = ________

5. There are 7 cars in the parking lot. I car drives away. How many cars are left? I.OA.6

$$\begin{array}{r} 7 \\ -\ 1 \\ \hline \end{array}$$

NAME DATE

Countdown: 16 Weeks

SCORE

THINK SMART FOR SBAC
On the actual test, you may be asked to select the correct answer using a mouse. In this book, you will circle the correct answer using a pencil.

1. Circle the number sentence that is false. 1.OA.6

$6 + 7 = 13$

$8 + 9 = 18$

$5 + 6 = 11$

2. Jamal has 10 pennies in his left hand. He has 3 pennies in his right hand. How many pennies does Jamal have in all? 1.OA.6

______ pennies

3. Use the number line to add. Write the sum. 1.OA.6

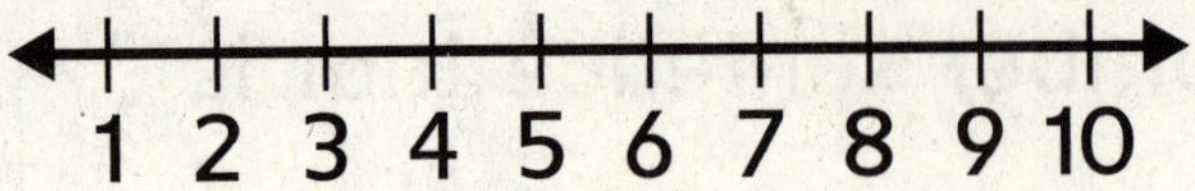

$$\begin{array}{r} 7 \\ +\ 3 \\ \hline \end{array}$$

4. Mr. Heng has 9 buckets. He buys 9 more. How many buckets does Mr. Heng have in all? 1.OA.6

_________ buckets

5. Jamie has 9 cookies. She gives 3 away. How many cookies does Jamie have left? 1.OA.7

_________ cookies

NAME ________ DATE ________

Countdown: 15 Weeks

SCORE ________

1. Tamika had 10 coins. She spent 4 of them. Tamika has 7 coins left. Circle true or false. 1.OA.7

   True   False

THINK SMART FOR SBAC
On the actual test, you may be asked to select true or false by clicking on the words. In this book, you will use a pencil to draw a circle around the correct word.

2. There are 8 kids playing. 2 kids go home for dinner. How many kids are still playing? Write a subtraction number sentence. Then write the related addition fact. 1.OA.6

   ______ − ______ = ______

   ______ + ______ = ______

3. Circle the addition sentence that is true. 1.OA.7

   $5 + 5 = 12$

   $8 + 8 = 16$

   $4 + 4 = 6$

4. Pedro has 6 red blocks and 2 yellow blocks. How many blocks does Pedro have in all? Write two ways to add. 1.OA.3

_________ + _________ = _________ blocks

_________ + _________ = _________ blocks

5. Jenny saw 2 grasshoppers in the grass. She saw 3 grasshoppers on the sidewalk. She saw 1 grasshopper on the porch. How many grasshoppers did Jenny see in all? 1.OA.2

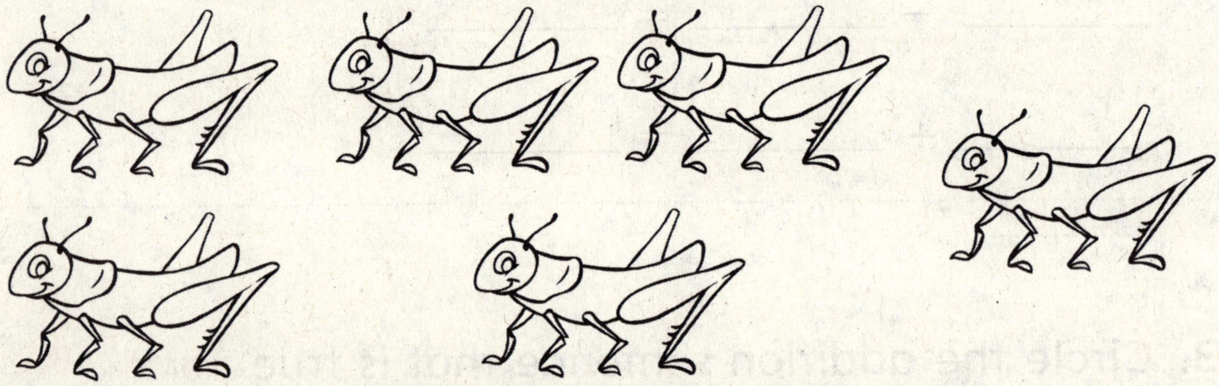

_________ grasshoppers

NAME ________ DATE ________

Countdown: 14 Weeks

SCORE ________

> **THINK SMART FOR SBAC**
> On the actual test, you may be asked to click into each blank to type the numbers. In this book, you will write the numbers using a pencil.

1. There are 11 robins in a tree. 3 of them fly away. How many robins are still in the tree? 1.OA.5

______ − ______ = ______ robins

2. There are 13 balloons. 3 balloons pop. How many balloons are left? Use the number line to help you subtract. 1.OA.5

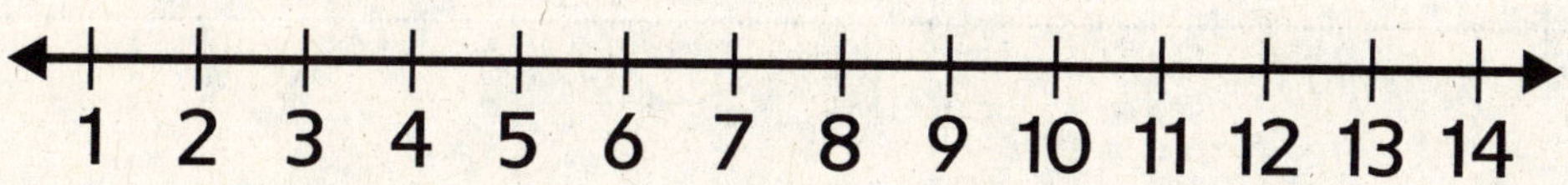

______ balloons

3. Add or subtract. Draw lines to match the doubles facts. 1.OA.6

$3 + 3 =$ ______ $8 - 4 =$ ______

$5 + 5 =$ ______ $6 - 3 =$ ______

$4 + 4 =$ ______ $10 - 5 =$ ______

4. Jasmine had 13 books. She gave 5 away. How many books does she have left? Write a number sentence to solve. 1.OA.6

______ ◯ ______ = ______ books

5. Winston has 5 rocks. He picks up 4 more. How many rocks does he have? 1.OA.6

$$\begin{array}{r} 5 \\ +\ 4 \\ \hline \end{array}$$

rocks

NAME ______ DATE ______

Countdown: 13 Weeks

SCORE ______

1. There are 17 frogs on a log. 9 of the frogs hop in the pond. How many frogs are left? Take apart the number to make 10. Then subtract. 1.OA.6

$17 - 9$

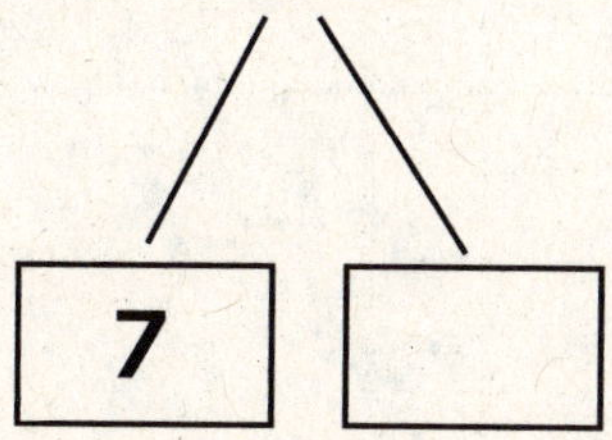

$17 -$ ______ $=$ ______

$10 -$ ______ $=$ ______

$17 - 9 =$ ______ frogs

2. Circle the number sentence that is not in the fact family. 1.OA.4

THINK SMART FOR SBAC
On the actual test, you may be asked to click on the facts to circle them. In this book, you will circle the answer using a pencil.

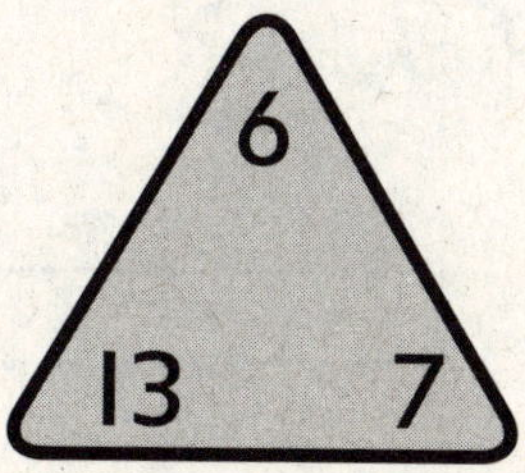

$6 + 13 = 7$

$13 - 6 = 7$

$6 + 7 = 13$

3. Andre sees 15 bugs. 6 of the bugs crawl away. How many bugs are left? Write an addition sentences and a subtraction sentence. 1.OA.6

_____ + _____ = _____

_____ − _____ = _____ bugs

4. Mr. Verne has 4 blacks hats and 2 brown hats. How many hats does Mr. Verne have in all? 1.OA.6

_____ hats

5. Khrystian has 9 oranges and 9 apples. Khrystian has 18 pieces of fruit. Circle true or false. 1.OA.7

true　　　false

NAME DATE

Countdown: 12 Weeks

SCORE

THINK SMART FOR SBAC
On the actual test, you may be asked to circle the objects by clicking with a mouse. In this book, you will circle them using a pencil.

1. Mara has the peanuts shown. Circle a group of ten peanuts. Write how many more. How many peanuts does Mara have in all? 1.NBT.2b

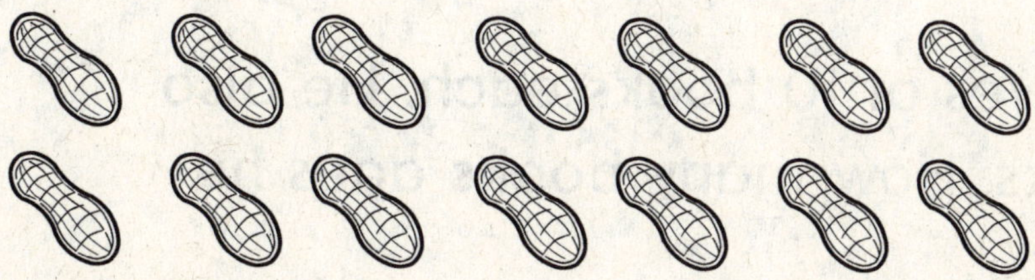

10 and ____ more ____ peanuts

2. Julia has 8 dimes. She spent 2 dimes. Count by tens. Write the numbers. How much does Julia have left? 1.NBT.1

____ ¢ ____ ¢ ____ ¢ ____ ¢ ____ ¢ ____ ¢ left

3. There are six boxes. Each box has 10 toy cars in it. How many cars are there in all? 1.NBT.4c

________ cars

4. Pedro has 7 shelves of 10 books each. He also has 5 more books. How many books does he have in all? 1.NBT.2a

________ books

5. Circle the number sentence that is not a doubles fact. 1.OA.7

$8 + 8 = 16$

$9 + 9 = 18$

$7 + 7 = 13$

NAME ____________ DATE ____________

SCORE ____________

Countdown: 11 Weeks

1. Lulu has 56 buttons. She has 5 sets of 10 green buttons. The rest are red. How many of the buttons are red? 1.NBT.2a

_______ red buttons

2. The Suarez family likes apples. They eat 10 apples every day. How many apples do they eat in 6 days. Fill in the table. 1.NBT.2c

THINK SMART FOR SBAC
On the actual test, you may be asked to click in the table cells and then type an answer. In this book, you will write your answers using a pencil.

Day	Apples

_______ apples

3. There are 4 groups of ten light bulbs. There are 3 extra light bulbs. How many light bulbs are there in all? 1.NBT.2a

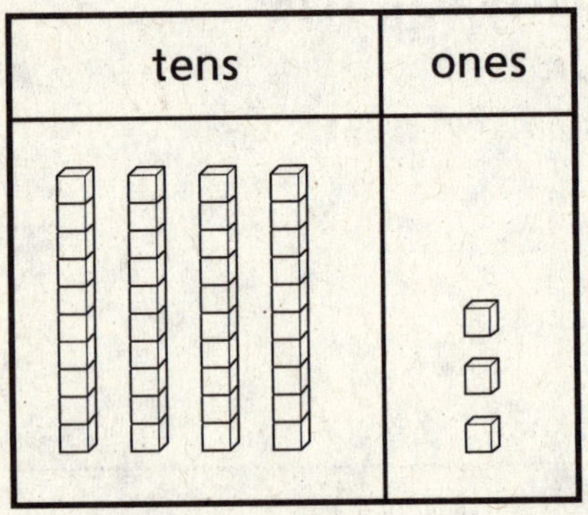

______ light bulbs

4. When I am added to 6, the sum is 11. What number am I? 1.OA.4

______

5. Henry found 3 small rocks. He also found 5 medium rocks and 4 big rocks. How many rocks did Henry find in all? 1.OA.2

______ + ______ + ______ = ______ rocks

NAME

DATE

Countdown: 10 Weeks

SCORE

1. Are the number sentences part of the fact family? Circle yes or no. I.OA.6

THINK SMART FOR SBAC
On the actual test, you may be asked to select yes or no by clicking. In this book, you will instead circle the answer using a pencil.

yes   no $7 + 5 = 12$

yes   no $5 + 7 = 12$

yes   no $7 - 5 = 12$

yes   no $12 - 5 = 7$

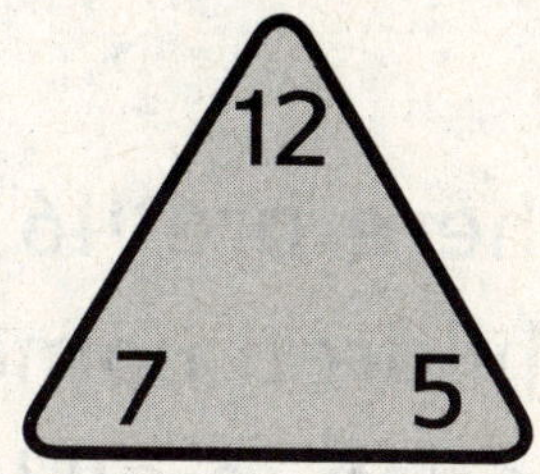

2. Alma's team has 36 points. The team scores 20 more points. How many points does the team have now? I.NBT.4

$$\begin{array}{r} 36 \\ +\ 20 \\ \hline \end{array}$$

points

3. There are 20 students in one class. There are 30 students in another class. How many students are there in all? I.NBT.4

________ students

4. There are 46 children at the beach. 3 more children come. How many children are at the beach in all? I.NBT.4

	tens	ones
+		

________ children

5. Mrs. Gomez has two kinds of flowers. She has 13 flowers in all. Circle the kinds of flowers Mrs. Gomez has. I.NBT.4

6 roses

4 daisies

7 tulips

NAME ________________ DATE ________

Countdown: 9 Weeks

SCORE ________

1. Mr. Garmen has 17 chairs. He brings in 8 more. How many chairs does he have in all? Circle the ones to show regrouping. Write your answer. 1.NBT.4

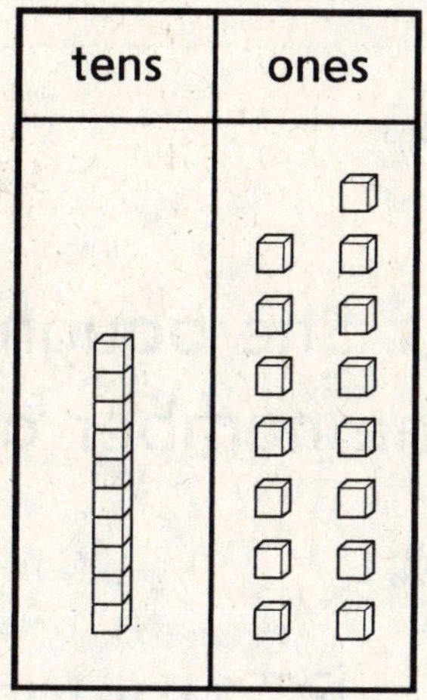

________ chairs

2. 40 + ________ = 80. Mark the related subtraction fact. 1.NBT.6

THINK SMART FOR SBAC
On the actual test, you may be asked to click into the boxes to select them. In this book, you will shade the boxes using a pencil instead.

- ☐ 40 + 10 = 80
- ☐ 80 − 40 = 40
- ☐ 80 − 40 = 20
- ☐ 40 + 30 = 70

3. There are 50 players on a football team. 20 players leave the team. How many players are left? 1.NBT.6

_________ players

4. Kylie had 47 stamps. She bought ten more stamps. Underline the number of stamps that Kylie has. 1.NBT.5

27 stamps   37 stamps

47 stamps   57 stamps

5. Quentin buys a cookie that costs 45¢. He only used nickels to pay. How many nickels did Quentin use? 1.NBT.1

_________ nickels

NAME .... DATE ....

Countdown: 8 Weeks

SCORE ....

1. Write the totals. How many people like the color blue? I.MD.4

Favorite Colors		
Color	**Tally**	**Total**
Blue	~~||||~~ ~~||||~~ ||	
Green	~~||||~~ |||	
Yellow	||||	

_________ people like blue.

2. George asks his friends to name their favorite sport. 4 people like football. 2 people like baseball. 3 people like soccer. Make a graph. I.MD.4

THINK SMART FOR SBAC
On the actual test, you may be asked to drag the pictures into the graph in order to tell how many. In this book, you will draw the pictures using a pencil.

Our Favorite Sports

Soccer								
Football								
Baseball								

3. Which kind of plant has more votes than flowers? 1.MD.4

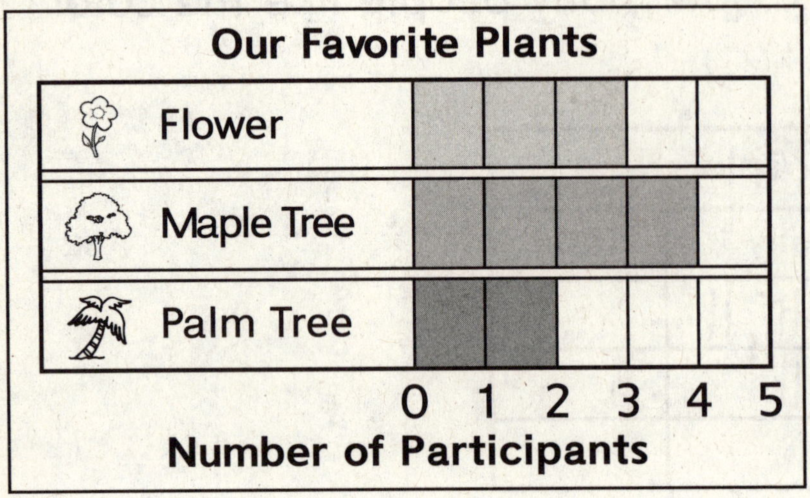

Answer: ____________________

4. Rayshawn has 7 small cars and 6 big cars. Rayshawn has 14 cars. Circle true or false. 1.OA.7

true   false

5. Write each number. Circle *is greater than, is less than,* or *is equal to*. 1.NBT.3

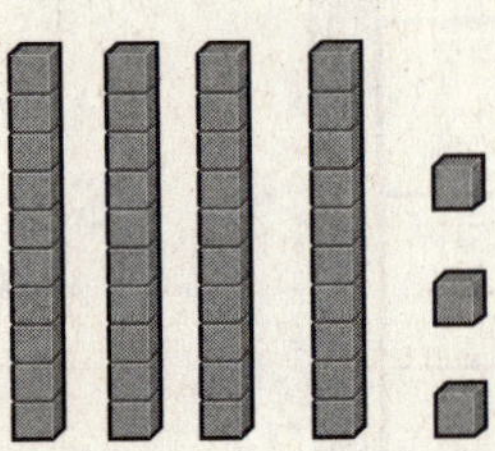

is greater than

is less than

is equal to

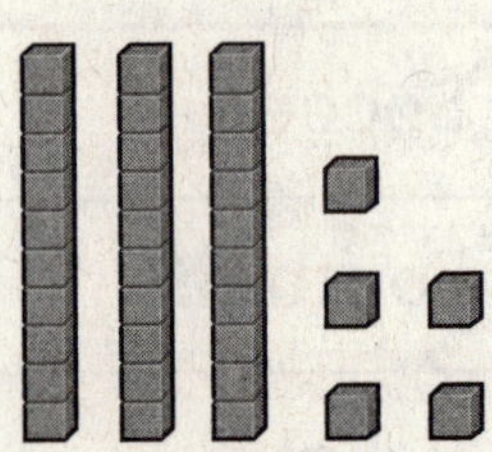

__________   __________

NAME .... DATE ....

Countdown: 7 Weeks

SCORE ....

1. Write the tallies. What flavor do people like the most? I.MD.4

> **THINK SMART FOR SBAC**
> On the actual test, you may be asked to drag the tally marks into the table using a mouse. In this book, you will draw the tally marks using a pencil.

Favorite Flavors		
Flavor	**Tally**	**Total**
Vanilla		6
Chocolate		7
Strawberry		9

People like ________________ the most.

2. Mercedes is counting toys. She counts 12 toys in all. How many dolls does she count? I.MD.4

________ dolls

3. Write the numbers in to make a fact family. 1.NBT.6

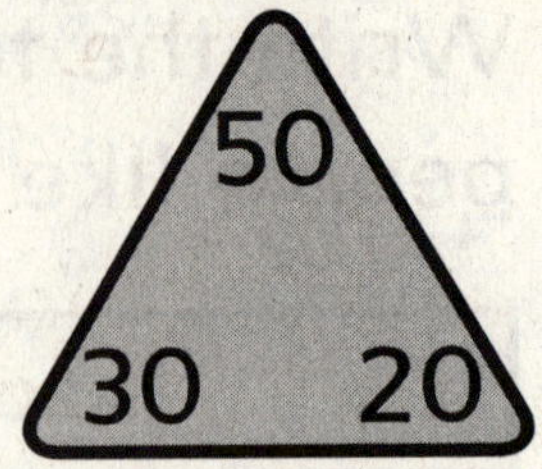

_____ + _____ = _____

_____ + _____ = _____

_____ − _____ = _____

_____ − _____ = _____

4. There are 12 fish near a rock. 2 of them swim away. How many fish are left? 1.OA.5

_____ − _____ = _____ fish

5. Jamal, Winston, and Medina each have a piece of fruit. The fruits are a lime, a lemon, and a plum. Jamal has a purple fruit. Medina has a green fruit. Winston has a yellow fruit. Make a table. 1.MD.4

Whose fruit is a lemon?		
Name	**Color**	**Fruit**
Jamal		
Winston		
Medina		

Whose fruit is a lemon? _____________

NAME .... DATE ....

Countdown: 6 Weeks

SCORE ....

1. A skateboard is shorter than a bike. A bike is shorter than a car. Is a car longer than or shorter than a skateboard? Circle the answer. 1.MD.1

shorter than   longer than

---

2. Naomi has three brushes. Match each brush to the number. 1 is for short. 2 is for shorter. 3 is for shortest. 1.MD.1

> **THINK SMART FOR SBAC**
> On the actual test, you may be asked to draw lines using a mouse. In this book, you will draw the lines using a pencil.

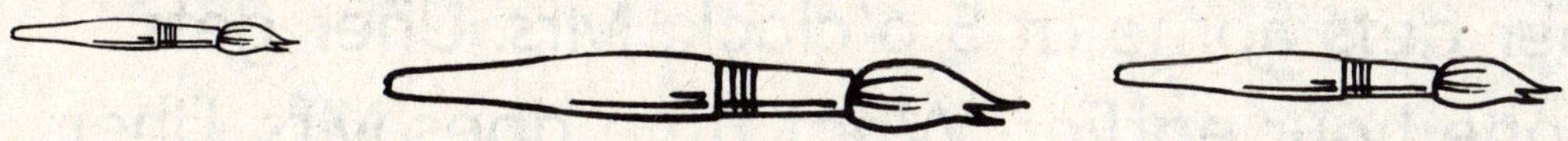

1   2   3

3. A pea plant was 4 cubes high on Monday. It grows 2 cubes higher every day. How many cubes high is it on Wednesday? Complete the table. I.MD.2

Days	Cubes
Monday	
Tuesday	
Wednesday	

______ cubes high

4. Mr. Uher gets home at 5 o'clock. Mrs. Uher gets home one hour earlier. What time does Mrs. Uher get home? I.MD.3

______ o'clock

5. There are 40 oranges in one basket. Geraldine puts 30 more oranges in. How many oranges are in the basket? I.NBT.4

______ oranges

NAME DATE

Countdown: 5 Weeks

SCORE

1. Kiane eats dinner at 6 o'clock. She starts reading one hour later. Mark the time on the clock that Kiane starts reading. I.MD.3

THINK SMART FOR SBAC
On the actual test, you may be asked to drag the hands around the clock to mark the correct time. In this book, you will draw the hands using a pencil.

2. Liam finished running at the time on the clock. He ran for 1 hour. What time did Liam start running? I.MD.3

________ o'clock

3. Yolanda's mom picks her up from school at 3:00. They drive 30 minutes home. What time do they get home? I.MD.3

____

4. Add or subtract. Draw lines to match the doubles facts. I.OA.6

7 + 7 = ________	14 − 7 = ________
9 + 9 = ________	12 − 6 = ________
6 + 6 = ________	18 − 9 = ________

5. A store has 50 toy cars. They sell 40 of the cars. How many toys cars are left? Use the number line to help. I.NBT.6

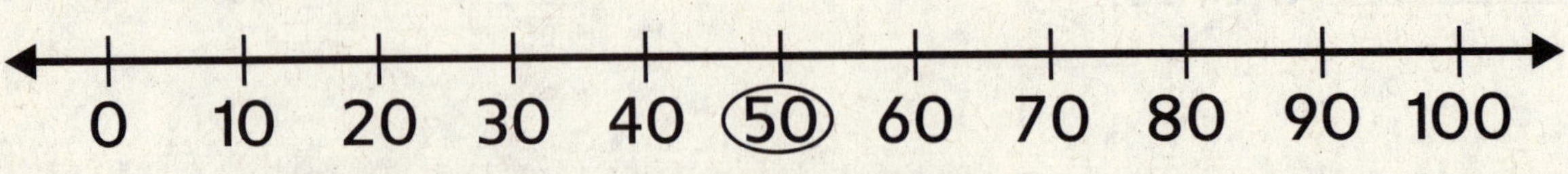

________ toy cars

NAME ____________ DATE ____________

Countdown: 4 Weeks

SCORE ____________

1. Circle the object that has a square. I.G.1

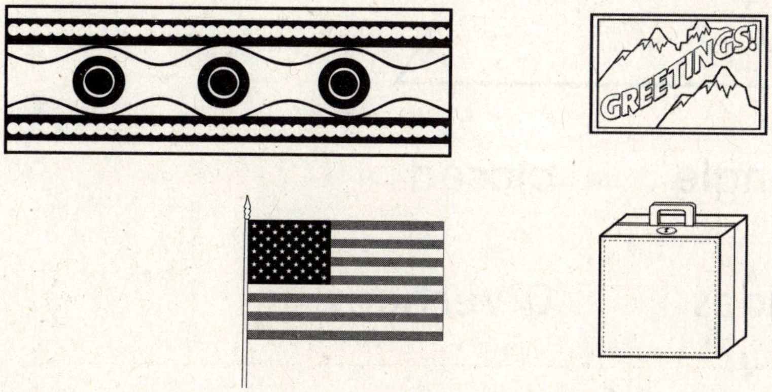

2. Draw a line in the picture to make a triangle and a trapezoid. I.G.1

THINK SMART FOR SBAC
On the actual test, you may be asked to draw the line using a mouse. In this book, you will draw the line using a pencil.

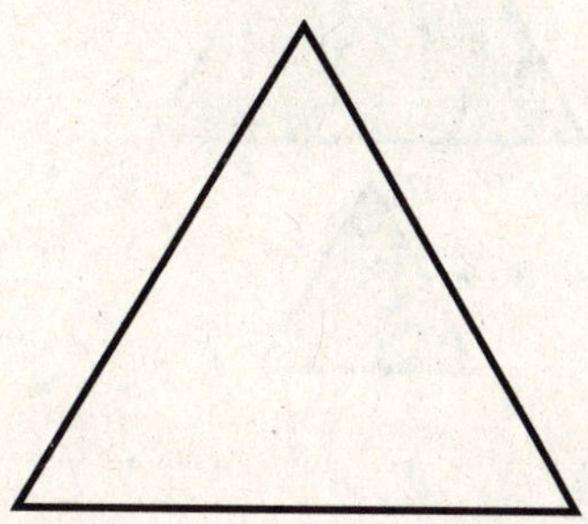

3. Circle the words that describe the shape. 1.G.1

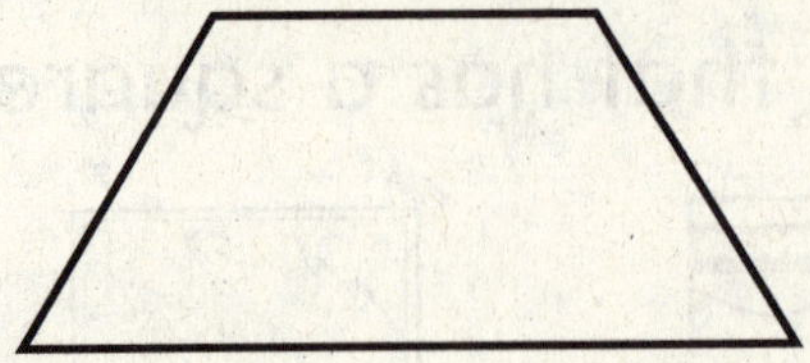

trapezoid   triangle   closed

3 vertices   4 sides   0 vertices

4. Francisco wants to cover the hexagon with 4 shapes. How many of each shape does he need to make the hexagon? 1.G.2

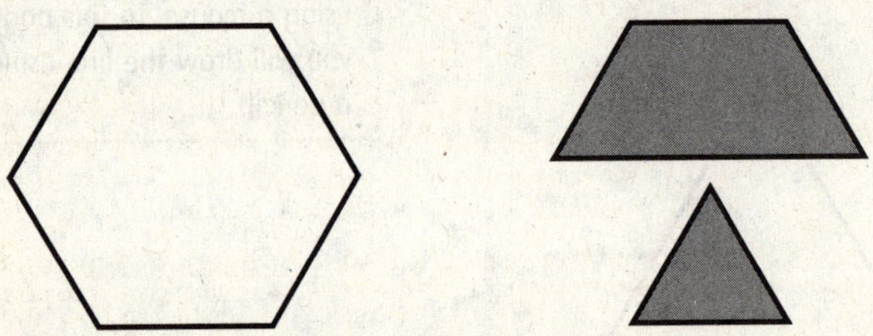

______

______

5. Josiah has practice at half past 3. Practice lasts for one hour. At what time is practice over? Write the time on the clock. 1.MD.3

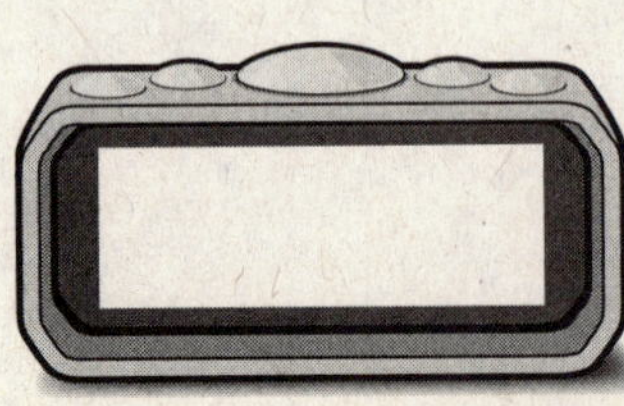

NAME DATE

Countdown: 3 Weeks

SCORE

1. Count how many of each shape. I.G.1

_______ trapezoids _______ circles

_______ rectangles _______ triangles

2. Circle the shapes that can be used to make the design. I.G.2

THINK SMART FOR SBAC
On the actual test, you may be asked to circle the shapes by clicking on them. In this book, you will draw the circles using a pencil.

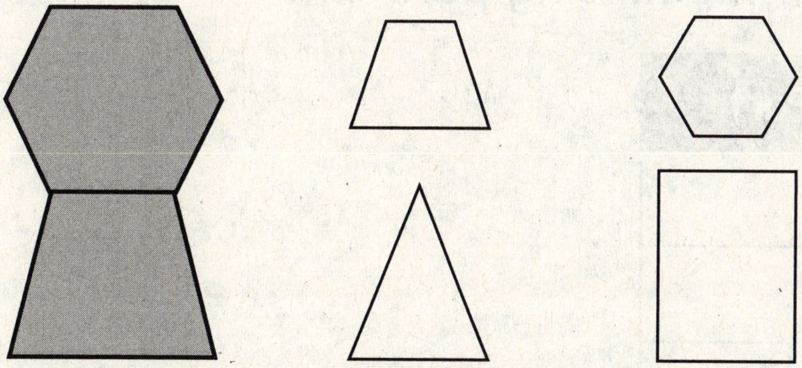

3. Mrs. Chin bakes a cake. She cuts it into 4 equal parts. Circle the pictures that could be Mrs. Chin's cake. I.G.3

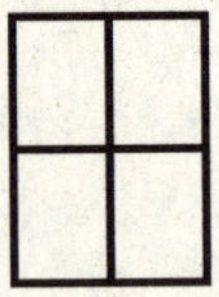

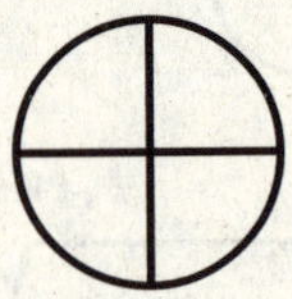

4. It is half past the hour. The hour hand is between 3 and 4. What time is it? Circle the correct time. I.MD.3

2:30 3:30 4:30

5. Subtract to fill in the missing part. I.OA.I

Part	Part
10	
Whole	
15	

NAME DATE

Countdown: 2 Weeks

SCORE

1. What is the name of the shape that makes up the faces of the object? I.G.1

____

2. Circle the names of the shapes that are not used to make the solid. I.G.2

THINK SMART FOR SBAC
On the actual test, you may be asked to click on the words to select them. In this book, you will draw the circles using a pencil.

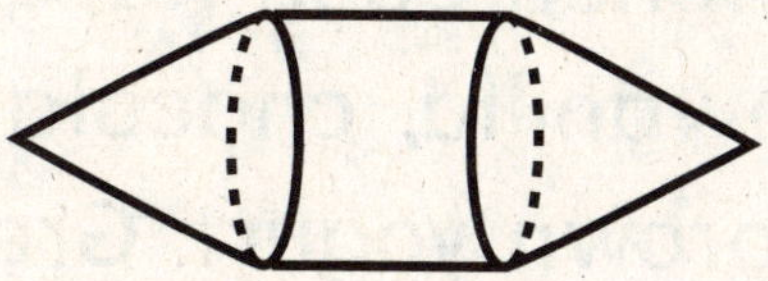

cone circle

cylinder rectangular prism

triangle

3. A fish tank has six rectangular faces. What shape is the fish tank? I.G.I

____

4. Mr. Fan has ice cream that is served in a cone. The cone has a face that is a square. Circle true or false. I.G.I

true false

5. Chance, Lamar, and Gretchen each have a cup of yogurt. The flavors are vanilla, chocolate, and strawberry. Chance has brown yogurt. Gretchen has white yogurt. Lamar has pink yogurt. Make a table. I.MD.4

Whose flavor is chocolate?

Name	Color	Flavor
Chance		
Lamar		
Gretchen		

Whose flavor is chocolate? ______________

NAME .... DATE ....

Countdown: 1 Week

SCORE ....

1. Garrett made a pattern with blocks. What two shapes does he need to complete his pattern? 1.G.1

cone   rectangular prism   cylinder

2. Underline the shapes that have six faces. 1.G.1

THINK SMART FOR SBAC
On the actual, test you may be asked to underline the words using a mouse. In this book, you will underline the words using a pencil.

cone   cylinder

cube   rectangular prism

3. Juan is counting fruit. 1.MD.4

How many more oranges than apples are there? ________

4. Isabella wants to divide a trapezoid into a rectangle and two triangles. Draw the lines to help Isabella. 1.G.2

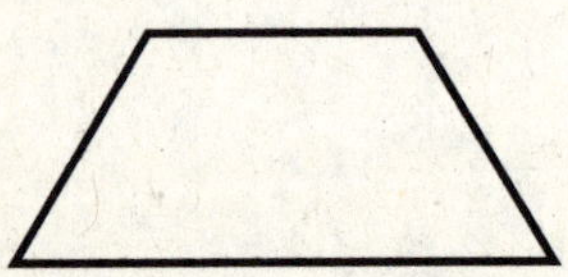

5. A can of soup has 2 faces shaped like circles and no vertices. What is the name of the shape? 1.G.1

____

NAME DATE

Chapter 1 Test

SCORE

1. Find the matching addition number sentence. 1.OA.1

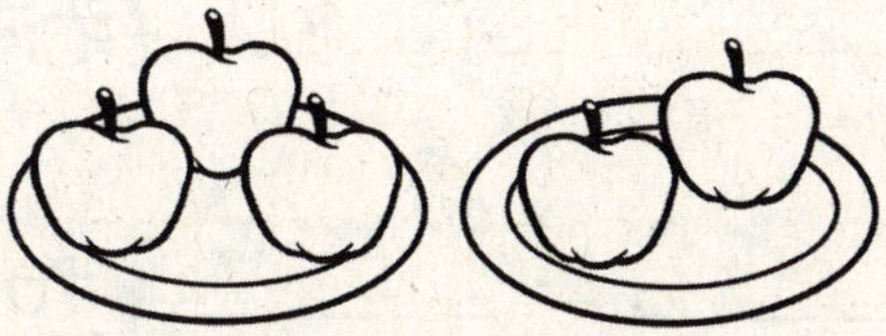

A. $3 + 2 = 5$

B. $2 + 3 = 6$

C. $1 + 4 = 5$

2. Find the missing part. 1.OA.1

Part	Part
7	2
Whole	

3. Write an addition number sentence. I.OA.1

________ + ________ = ________ turtles

4. Add zero. I.OA.3

9 + ________ = ________

5. True or false?

Jeremiah has 5 toy cars. He gets 2 more for his birthday. Jeremiah has 6 cars in all. I.OA.1, I.OA.7

True False

6. True or false?

Fernando has 2 books on one shelf and 3 books on another shelf. Fernando has 6 books in all. I.OA.6, I.OA.7

True   False

7. Add. I.OA.6

$4 + 3 = $ ______

8. Add. I.OA.6

$5 + 1 = $ ______

9. True or false? 1.OA.6, 1.OA.7

$2 + 4 = 6$

True False

10. Add. 1.OA.6

$$\begin{array}{r} 5 \\ +\ 5 \\ \hline \end{array}$$

11. Add. 1.OA.6

$$\begin{array}{r} 1 \\ +\ 8 \\ \hline \end{array}$$

12. Find the missing part. 1.OA.6

Part	Part
6	
Whole	
10	

13. Geraldine has 5 fish. Circle the ways that will make 5 fish. 1.OA.6

2 + 3   4 + 2

1 + 4   0 + 5

14. Mr. Bake has 4 brown chairs and 4 black chairs. How many chairs does Mr. Bake have in all? 1.OA.6

_________ chairs

15. Circle two numbers that will make 10. 1.OA.6

2	6	3
0	8	5

NAME DATE

Chapter 2 Test

SCORE

1. Find the matching subtraction number sentence. 1.OA.1, 1.OA.5

A. 8 − 5 = 3

B. 9 − 6 = 3

C. 9 − 3 = 6

2. Find the missing part. 1.OA.4, 1.OA.8

Part	Part
2	
Whole	
6	

Subtract. I.OA.6

3. 5 − 2 = ________

4. 10 − 2 = ________

5. True or false? I.OA.I, I.OA.7

Zoe has 6 books. She takes 2 back to the library. Now Zoe has 5 books.

True　　　　False

6. Write a subtraction number sentence. 1.OA.1

________ − ________ = ________

7. Subtract 0. 1.OA.6, 1.OA.8

5 − ________ = ________

8. True or false? 1.OA.7

4 − 3 = 1

True False

Subtract. 1.OA.6

9. $\begin{array}{r} 10 \\ -\ 5 \\ \hline \end{array}$

10. $\begin{array}{r} 7 \\ -\ 6 \\ \hline \end{array}$

11. $\begin{array}{r} 2 \\ -\ 0 \\ \hline \end{array}$

12. Write the related subtraction fact. 1.OA.6

$$8 - 3 = 5$$

________ − ________ = ________

13. Find the missing part. 1.OA.4, 1.OA.8

Part	Part
6	
Whole	
10	

A. 4

B. 3

C. 5

14. Write a subtraction number sentence. 1.OA.1

Xavier has 5 cheese cubes. He eats 3 of them. How many cheese cubes are left?

________ − ________ = ________

15. Find the related subtraction number sentence. 1.OA.6

$$10 - 3 = 7$$

A. $7 - 4 = 3$

B. $10 - 6 = 4$

C. $10 - 7 = 3$

NAME DATE

SCORE

Chapter 3 Test

1. Chase starts at the number 9. He counts "9, 10, 11, 12." Circle the number sentence that matches Chase's counting. 1.OA.5

9 + 3 = 12

10 + 2 = 12

9 + 4 = 12

2. Xavier has 7 pennies in his left pocket and 6 pennies in his right pocket. How many pennies does Xavier have in all? 1.OA.6

7¢ + 6¢ = ______ ¢

3. Use the number line to add. Write the sum. 1.OA.6

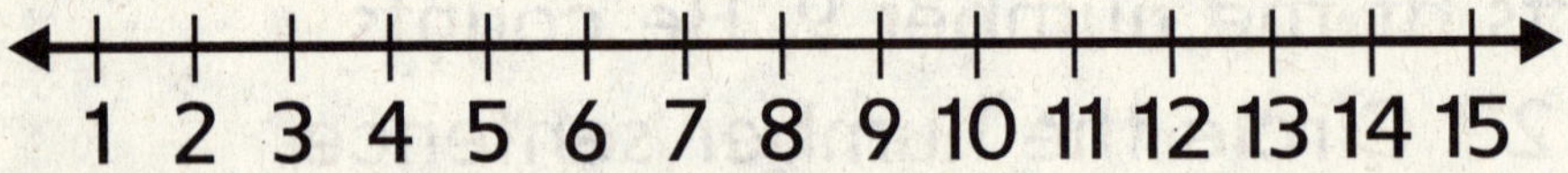

$$\begin{array}{r} 5 \\ +\ 9 \\ \hline \end{array}$$

4. Circle the correct doubles facts. 1.OA.6

5 + 5 = 12 6 + 6 = 10

4 + 4 = 8 7 + 7 = 14

5. Shalyn took 8 pictures on Friday. She took 5 more pictures on Saturday. How many pictures did she take in all? 1.OA.6

________ pictures

6. Javier's teacher asked him to add 8 + 4. He makes 10 to add. Which number sentence does Javier use? 1.OA.6

10 + 1 = 11

10 + 2 = 12

10 + 3 = 13

7. Darian has 5 baseball cards and 9 football cars. How many cards does Darian have in all? Write two ways to add. 1.OA.3

________ + ________ = ________ cards

________ + ________ = ________ cards

8. Christina bought 3 apples, 5 oranges, and 9 bananas. How many pieces of fruit did she buy in all? 1.OA.2

_____ + _____ + _____ = _____ pieces

9. True or false? 1.OA.2, 1.OA.7

4 + 3 + 2 = 10

True False

10. Which sum equals 20? 1.OA.2

A. 5 + 6 + 7

B. 2 + 7 + 8

C. 6 + 6 + 8

11. True or false?

Bryn has 6 white flowers and 7 red flowers. Her sister has 7 blue flowers and 6 pink flowers. Bryn and her sister have the same number of flowers. 1.OA.3, 1.OA.7

True False

12. Make 10 to add $8 + 6$. 1.OA.6

$\_\_\_\_\_ + \_\_\_\_\_ = \_\_\_\_\_$

13. Circle any sums that equal 15. 1.OA.6

10 + 5   9 + 6

8 + 7   6 + 4

14. Veronica saw 9 birds in her front yard and 9 birds in her back yard. How many birds did she see in all? 1.OA.6

________ birds

15. Kerin's team scored 8 points. The team scored 6 more points. How many points did they have in all? 1.OA.6

________ points

NAME ________ DATE ________

Chapter 4 Test

SCORE ________

1. Derrick starts counting backwards from 19. He says, "19, 18, 17, 16." Circle the number sentence he is trying to solve. 1.OA.5

$19 - 3 = 16$

$19 - 4 = 16$

$19 - 4 = 15$

2. There are 12 muffins. The Perez family eats 5 of them. How many muffins are left? Use the number line to help you subtract. 1.OA.5

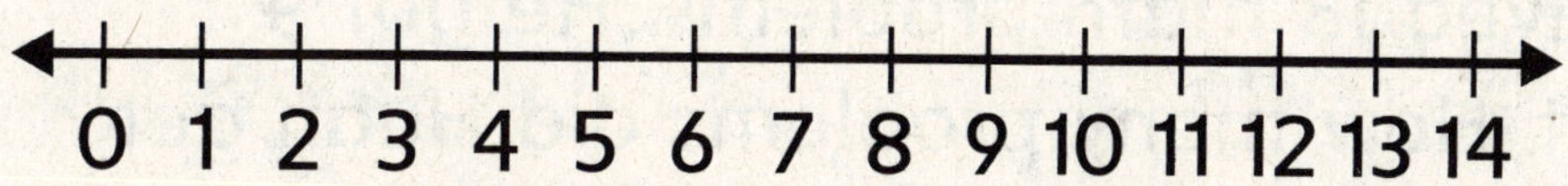

________ muffins

3. Add or subtract. Draw lines to match the doubles facts. 1.OA.6

6 + 6 = ______	14 − 7 = ______
7 + 7 = ______	12 − 6 = ______
8 + 8 = ______	16 − 8 = ______

4. There are 18 bats in the zoo. 9 are sleeping. How many bats are awake? 1.OA.6

______ bats

5. Huan solved 13 math problems. He got 4 incorrect. How many problems did Huan get correct? Write a number sentence to solve. 1.OA.1

______ ◯ ______ = ______ problems

6. Write a subtraction sentence for the following number line. 1.OA.5

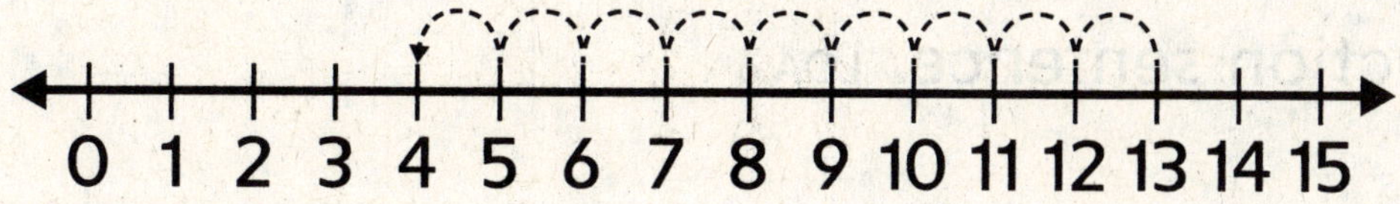

________ − ________ = ________

7. There are 15 elephants drinking at the spring. 7 elephants leave. How many elephants are left? Take apart the number to make 10. Then subtract. 1.OA.6

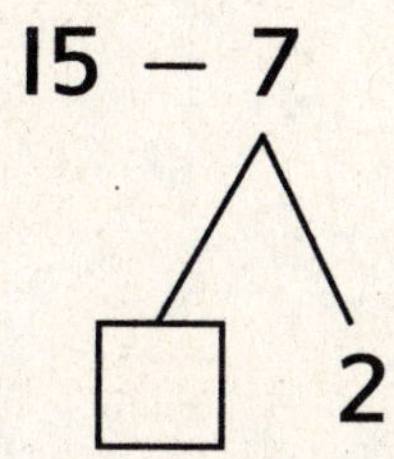

15 − ________ = ________

10 − ________ = ________

15 − 7 = ________ elephants

8. Iesha checked out 14 books from the library. She read 8 of the books. How many books did she not read? Write an addition sentence and a subtraction sentence. 1.OA.6

_______ + _______ = _______

_______ − _______ = _______ books

9. Which subtraction fact is not related to $5 + 6 = 11$? 1.OA.6

A. $11 - 5 = 6$

B. $11 - 6 = 5$

C. $6 - 11 = 5$

10. Circle the number sentence that is not in the fact family. 1.OA.4

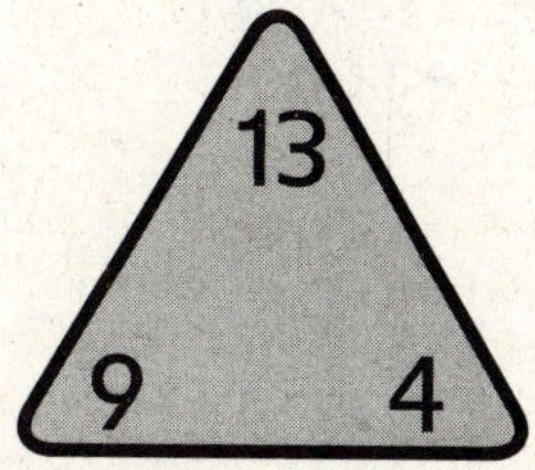

$4 + 9 = 13$

$9 + 13 = 4$

$13 - 9 = 4$

11. Maria is thinking of a number. When she adds 6 to the number, she gets 10. What is her number? 1.OA.4

6 + ________ = 10

12. Marquis has 6 shoes and some socks. He has 14 shoes and socks in all. How many socks does Marquis have? 1.OA.4

________ socks

13. True or false?

The two number sentences are in the same fact family. I.OA.4

$9 + 9 = 18$

$18 - 9 = 9$

True False

14. Which subtraction problem has the same answer as $12 - 5$? I.OA.6

$10 - 2$

$10 - 3$

$10 - 4$

15. Lena listened to 15 songs. She liked 10 of the songs. How many songs did Lena not like? I.OA.4

$10 +$ ________ $= 15$

________ songs

NAME ____________ DATE ____________

Chapter 5 Test

SCORE ____________

1. Mrs. Browning has the scissors shown. Circle a group of ten scissors. Write how many more. How many scissors does Mrs. Browning have in all? I.NBT.2b

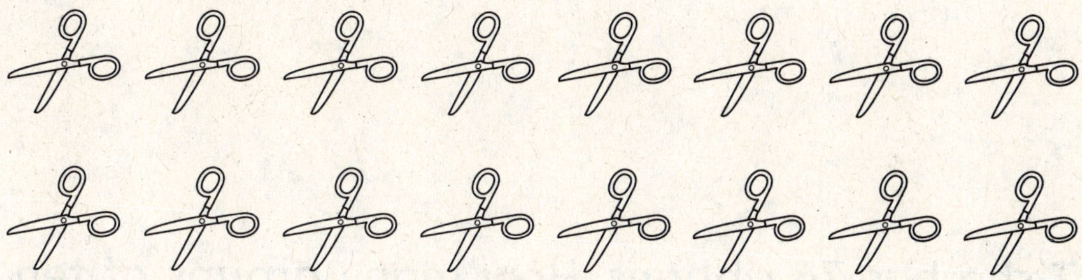

10 and ________ more ________ scissors

2. Gwen has 10 white flowers and 8 red flowers. How many flowers does she have in all? I.NBT.2c

________ flowers

3. Eunice has 40 baskets. She gets 2 more. How many baskets does Eunice have? I.NBT.2a

A. 42 baskets

B. 24 baskets

C. 44 baskets

4. Tisha has 76 ribbons. How many groups of ten does she have? How many ones? I.NBT.2a

________ tens and ________ ones

5. True or false? George has 56 marbles. Samuel has 65 marbles. Samuel has ten more marbles than George. I.NBT.5

True False

6. Jackson had a lemonade stand. Count by tens to find out how much money Jackson made. I.NBT.I

____ ¢ ____ ¢ ____ ¢ ____ ¢ ____ ¢ ____ ¢ ____ ¢ in all

7. There are 5 groups of ten cars. There are 2 extra cars. How many cars are there in all? I.NBT.2a

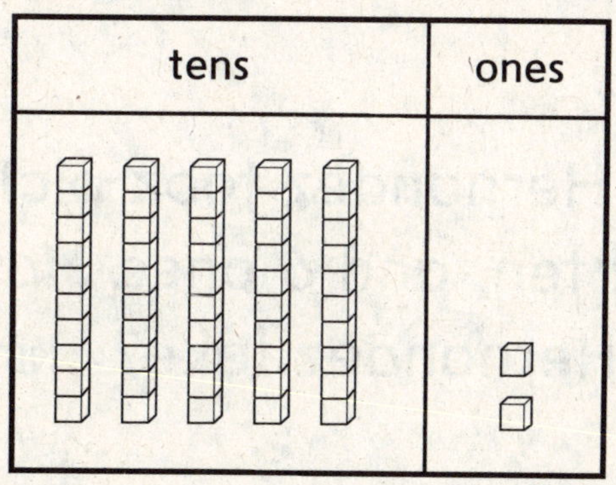

________ cars

8. Jorge buys a bracelet that costs 75¢. He only used nickels to pay. How many nickels did Jorge use? 1.NBT.1

________ nickels

9. Circle the right words. 1.NBT.3

is less than

46 is greater than 64

is equal to

10. On vacation Ms. Hernandez took pictures. She took 1 hundred, 1 ten, and 6 ones. How many pictures did Ms. Hernandez take? 1.NBT.1

________ pictures

11. Mr. Akita buys 10 stamps every week. How many stamps does he buy in 5 weeks? Fill in the table. 1.NBT.2c

Week	Stamps

______ stamps

12. Tory is thinking of a number. It is between 109 and 111. What is her number? 1.NBT.1

A. 108

B. 110

C. 112

13. Seamus has 35 nickels. Miranda has 41 nickels. Write <, >, or =. Tell who has more. I.NBT.3

35 ◯ 41

Who has more? ___________

14. Naomi picked eighty-seven apples. Circle the number of apples Naomi picked? I.OA.7

87

78

18

15. True or false?

7 tens and 8 ones is the same as 8 tens and 7 ones. I.NBT.5

True False

NAME ______ DATE ______

Chapter 6 Test

SCORE ______

1. There are 50 players on one team. There are 30 players on another team. How many players are there in all? 1.NBT.4

________ players

2. Circle the correct answer. 1.NBT.4

$$\begin{array}{r} 40 \\ +\ 40 \\ \hline \end{array}$$

A. 8

B. 48

C. 80

3. Gina sells 24 glasses of lemonade. She sells 20 more glasses. How many glasses does Gina sell in all? 1.NBT.4

$$\begin{array}{r} 24 \\ +\ 20 \\ \hline \end{array}$$

______ glasses

4. Add. 1.NBT.4

$$\begin{array}{r} 45 \\ +\ \ 3 \\ \hline \end{array}$$

5. Add. 1.NBT.4

$$\begin{array}{r} 5 \\ +\ 62 \\ \hline \end{array}$$

6. Sammy saw two kinds of bugs. He saw 16 bugs in all. Circle the kinds of bugs Sammy saw. 1.NBT.4

5 ladybugs

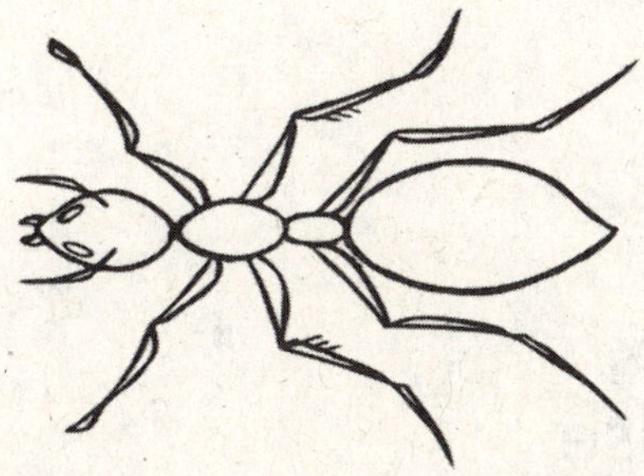

11 ants

6 beetles

7. Mr. Shimp has 26 pens. He buys 7 more. How many pens does he have in all? Circle the ones to show regrouping. Write your answer. 1.NBT.4

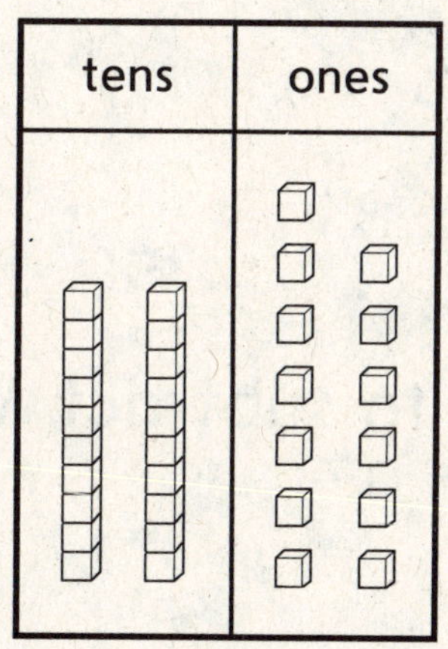

_______ pens

8. True or false? I.NBT.4

$$\begin{array}{r} 45 \\ +\ \ 7 \\ \hline 412 \end{array}$$

True   False

9. Kevin counted some cars. He counted 60 cars in all. 20 cars were red. The others were blue. How many blue cars did Kevin count? I.NBT.6

_______ blue cars

10. Use the number line to subtract. Write the difference. I.NBT.6

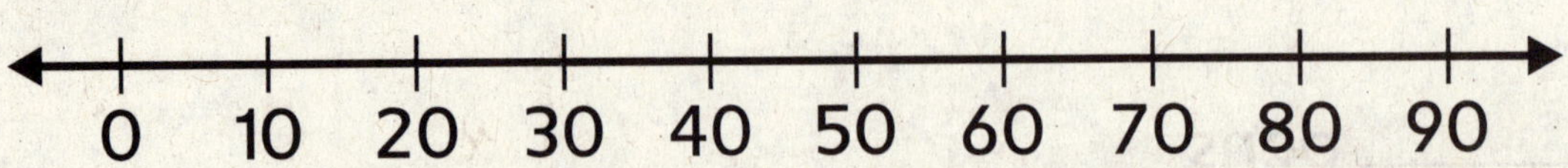

80 − 20 = _______

11. 30 + ______ = 70. Mark the related subtraction fact. I.NBT.6

☐ 70 − 30 = 40

☐ 70 − 30 = 10

☐ 40 + 30 = 70

12. Circle the number sentence that matches the number line. I.NBT.6

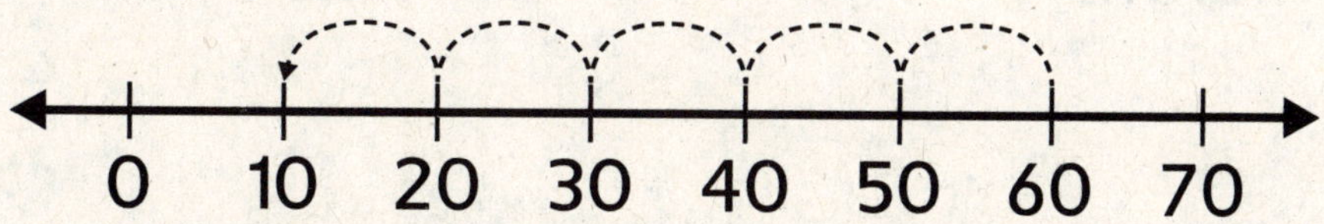

A. 60 − 50 = 10

B. 70 − 50 = 10

C. 60 − 50 = 20

13. Are the number sentences part of the fact family? Circle yes or no. 1.NBT.6

yes	no	30 + 30 = 60
yes	no	60 − 30 = 30
yes	no	6 − 3 = 3

14. Aaron hit 23 homeruns in his first season. In his second season he hit 20 homeruns. How many homeruns did Aaron hit in all? 1.NBT.4

________ homeruns

15. True or false?

76 is 70 more than 6. 1.NBT.4

True False

NAME ______________________ DATE ______________

Chapter 7 Test

SCORE ______________

1. Write the totals. I.MD.4

Favorite Season		
Season	**Tally**	**Total**
Winter	𝍸 ||||	
Spring	𝍸 𝍸	
Summer	||	
Autumn	𝍸 𝍸 |	

How many more people like Winter than Summer?

________ people

2. Henry, Chloe, and Jose each have a pet. The pets are a dog, a cat, and a bird. Henry has the pet that barks. Jose has the pet that tweets. Make a table. I.MD.4

Whose pet is a cat?		
Name	Sound	Pet

Whose pet is the cat? ________

3. Write the tallies. What sport do people like the most? I.MD.4

Favorite Sports		
Sport	**Tally**	**Total**
Football		7
Soccer		10
Tennis		5

People like ________ the most.

4. Arianna asks her friends to name their favorite donut. 7 people like plain. 3 people like iced. 2 people like filled. Make a graph. I.MD.4

Favorite Donut

Plain							
Iced							
Filled							

5. How many animals are there in all? I.MD.4

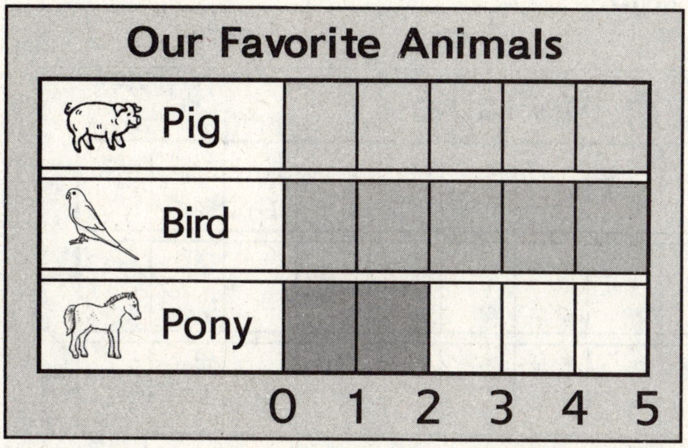

_________ animals

6. Mr. Johnson, Ms. Baird, and Mrs. Blackthorne each teach a math class. The sizes of the classes are 9, 15, and 22. Mrs. Blackthorne has the class with fewer than 10 students. Mr. Johnson has the class with more than 20 students. Make a table. I.MD.4

Whose class has 15 students?	
Name	How many students?

Whose class has 15 students? _______________

7. Mrs. Tyler has 5 plates, 8 forks, and 7 spoons. Make a bar graph. 1.MD.4

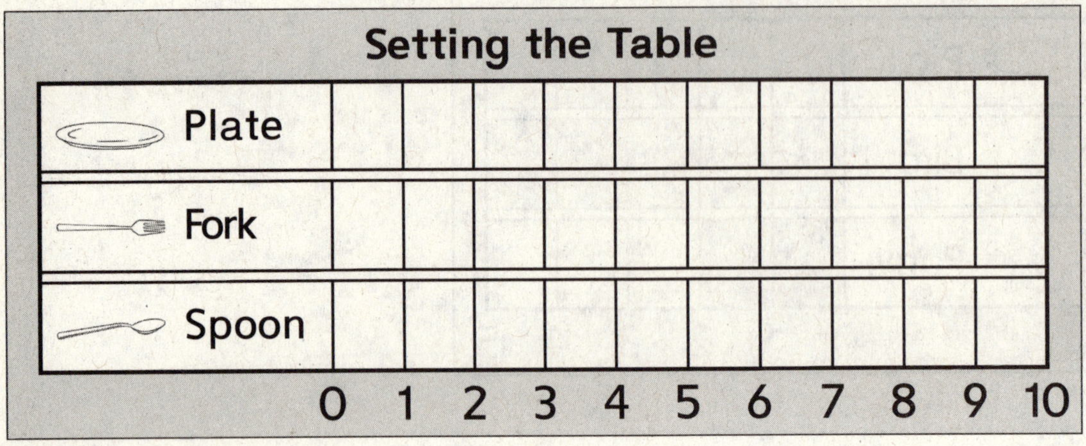

8. Look at the bar graph for the number of people who like each holiday. Make a tally chart. 1.MD.4

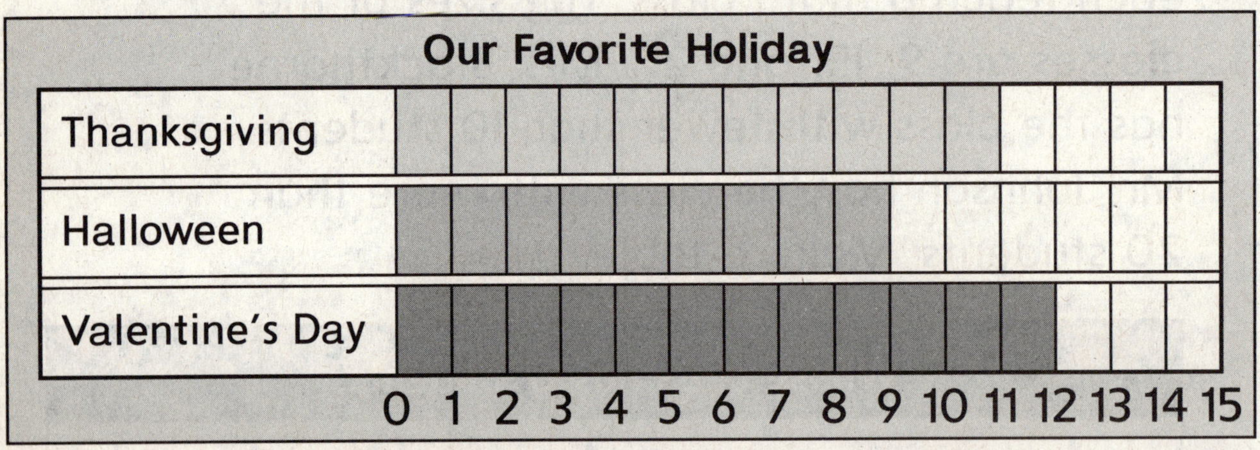

Favorite Holiday

Holiday	Tally	Total
Thanksgiving		
Halloween		
Valentine's Day		

9. Two teams scores 27 points in all. The Blue Jays score 14 points. Make a tally chart to show how many points each team scores. 1.MD.4

Points Scored		
Points	**Tally**	**Total**
Blue Jays		
Cardinals		

10. Francine is counting animals in a pond. She counts 15 animals in all. How many ducks does she count? 1.MD.4

_______ ducks

11. Camden collected coins during a lemonade sale. How many more dimes did he collect than pennies? 1.MD.4

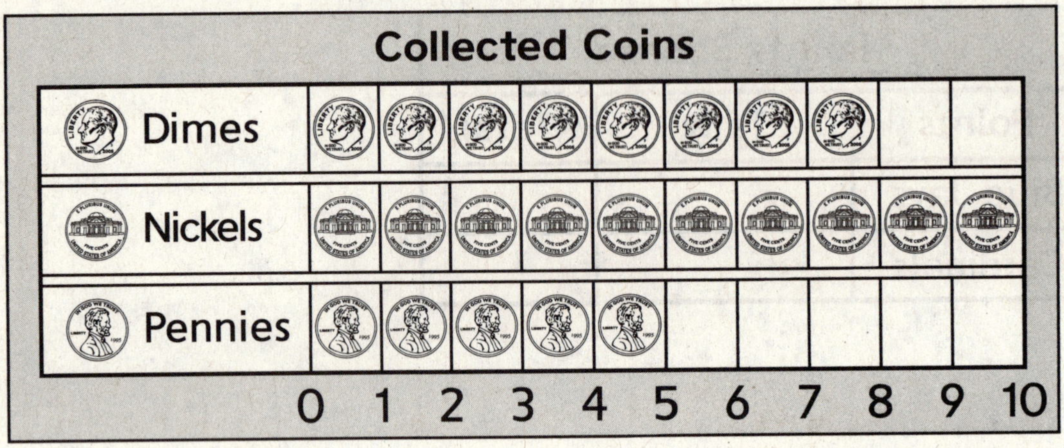

______ more dimes than pennies

12. Which kind of hat has more votes than the baseball cap? 1.MD.4

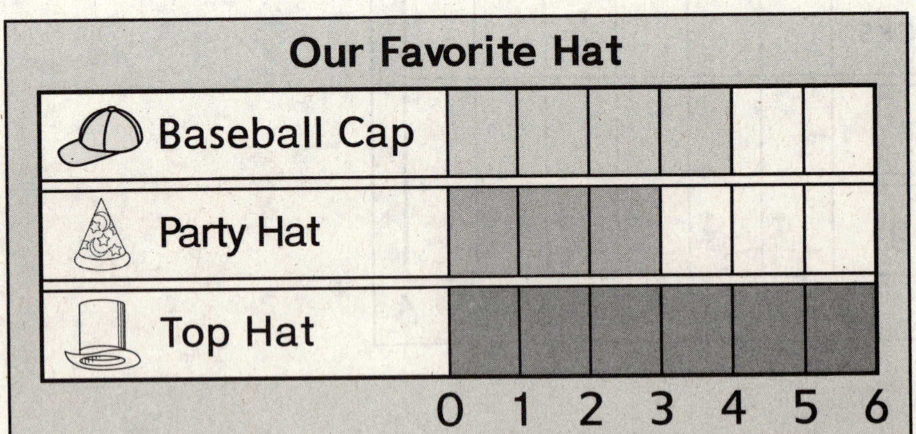

Answer: ______

NAME

DATE

Chapter 8 Test

SCORE

1. The table is longer than the chair. The carpet is longer than the table. Is the chair longer than or shorter than the carpet? Circle the answer. I.MD.I

shorter than   longer than

2. A pilot sees three planes. Match each plane to the number. I is for long. 2 is for longer. 3 is for longest. I.MD.I

I   2   3

3. How many acorns long is the caterpillar? I.MD.2

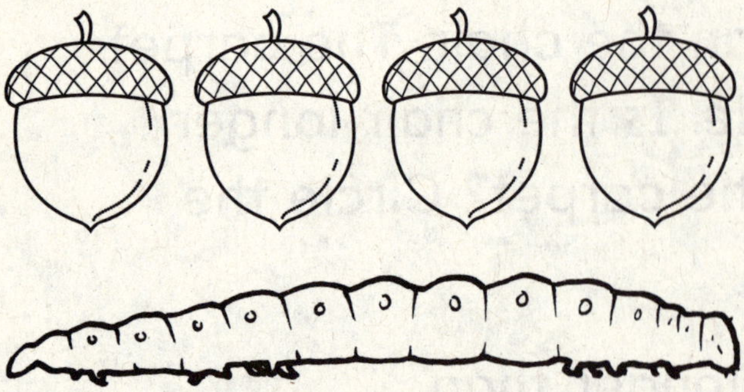

__________ acorns

4. A corn plant was 7 cubes high on Thursday. It grows 3 cubes higher every day. How many cubes high is it on Sunday? I.MD.2

Days	Cubes
Thursday	
Friday	
Saturday	
Sunday	

__________ cubes high

5. Match the times on the clocks. I.MD.3

6. It is half past the hour. The hour hand is between 7 and 6. What time is it? Circle the correct time. I.MD.3

5:30　　6:30　　7:30

7. Harry started his homework at 3 o'clock. He got home from school 1 hour earlier. What time did Harry get home? 1.MD.3

____________ o'clock

8. Pierre's mom drives to the mall at 4:30. She drives 30 minutes. What time does she arrive at the mall? 1.MD.3

____

9. The pencil is shorter than the piece of paper. The piece of paper is shorter than the book. Which object is the longest? 1.MD.1

A. The pencil

B. The piece of paper

C. The book

10. Amal had 17 cookies. He eats two cookies a day. How many cookies does he have left on Day 4? 1.MD.2

Days	Cookies
Day 1	17
Day 2	
Day 3	
Day 4	

__________ cookies

11. Geraldine wakes up at 7 o'clock. She eats breakfast one hour later. Mark the time on the clock that Geraldine starts eating breakfast. 1.MD.3

12. The Ramirez family leaves for the fair at the time shown on the clock. They arrive 2 hours later. Mark the time on the other clock that shows what time they arrived. I.MD.3

13. Martha has practice at half past noon. Practice lasts for one hour. What time is practice over? I.MD.3

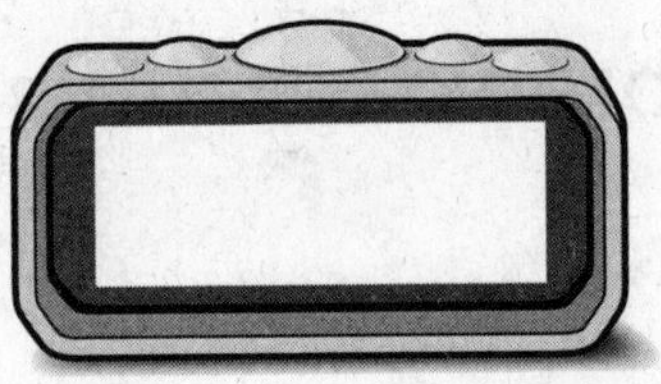

14. Circle the clock that shows the latest time. I.MD.3

NAME .... DATE ....

Chapter 9 Test

SCORE ....

1. Count how many of each shape. I.G.I

__________ trapezoids __________ circles

__________ rectangles __________ squares

2. Circle the shapes that can be used to make the design. I.G.2

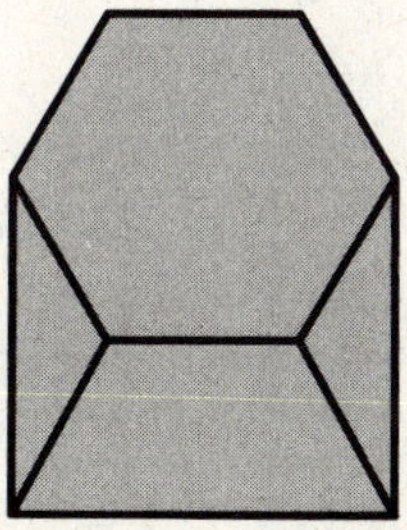

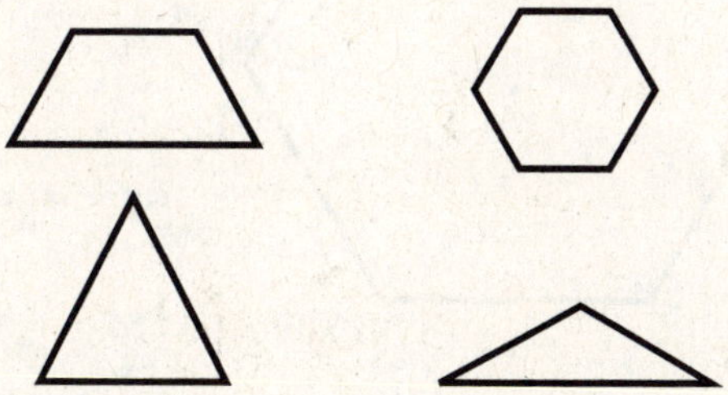

3. Circle all the words that describe the shape. I.G.1

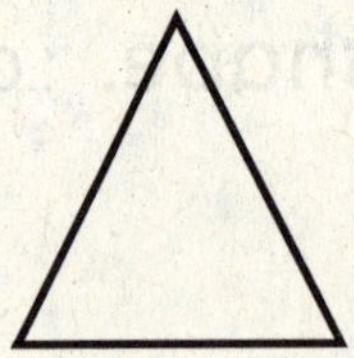

trapezoid triangle closed

3 vertices 4 sides 0 vertices

4. Draw two lines in the picture to make a rectangle and two triangles. I.G.1

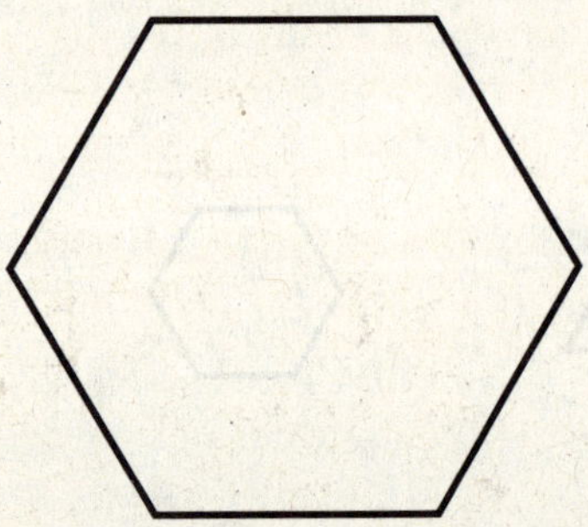

5. Circle the object that has a square in it. I.G.1

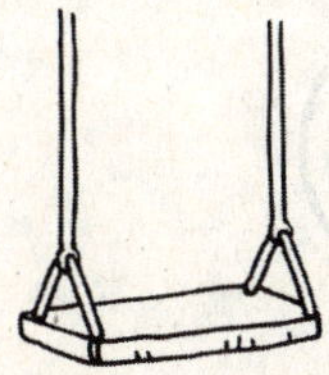

6. Thomas wants to cover the pentagon with 3 shapes. How many of each shape does he need to make the pentagon? I.G.2

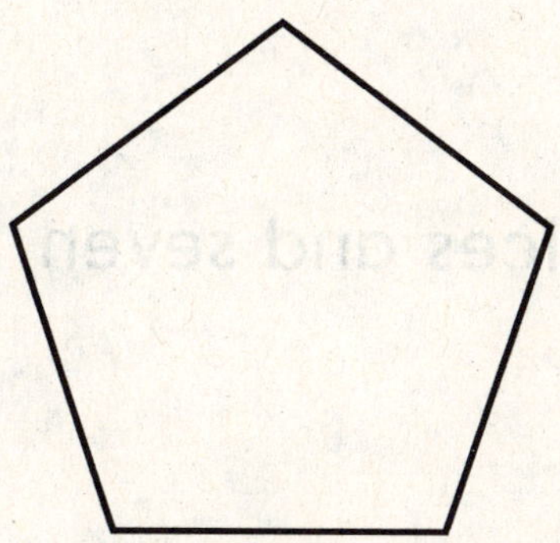

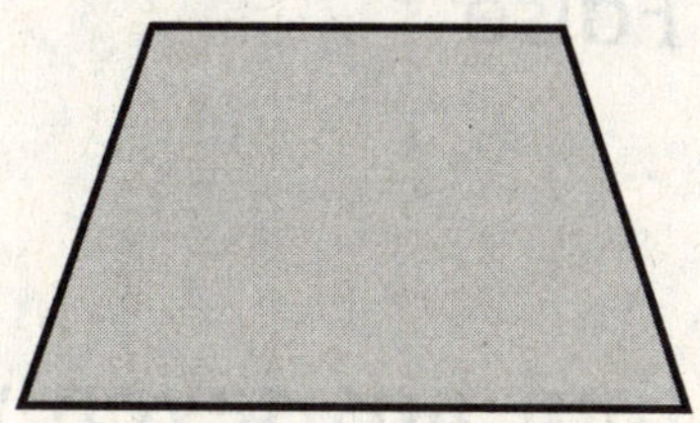

______

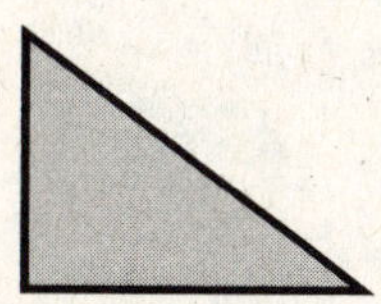

______

7. A pizza chef makes a pizza. He cuts it into 4 equal parts. Circle the pictures that could be the pizza. I.G.3

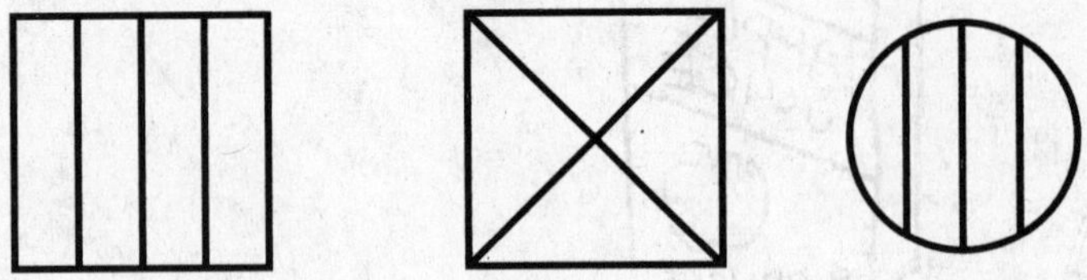

8. True or false?

A trapezoid has 4 vertices and 4 sides. I.G.I

True False

9. Draw a shape that has seven vertices and seven sides. I.G.I

10. Lamar is thinking of a shape with no vertices. What is the shape? I.G.1

A. square

B. circle

C. triangle

11. Enrique is building a table. He splits a wooden board into half. How many equal parts does he have?

__________ equal parts

12. Put an X on the shape that could not be used to cover the rectangle. I.G.1

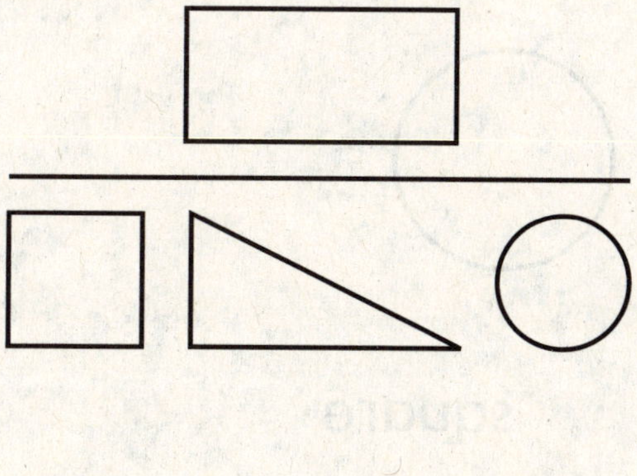

13. Circle the shape that shows thirds. I.G.3

14. Veronica cuts a piece of thread into exact fourths. How many pieces of thread does she have? I.G.3

A. 4

B. 14

C. 40

15. Put an X through any words that do not describe the shape. I.G.1

circle | square

0 vertices | 3 edges

NAME ____________ DATE ____________

Chapter 10 Test

SCORE ____________

1. Circle all shapes that make up the faces of the sponge. 1.G.1

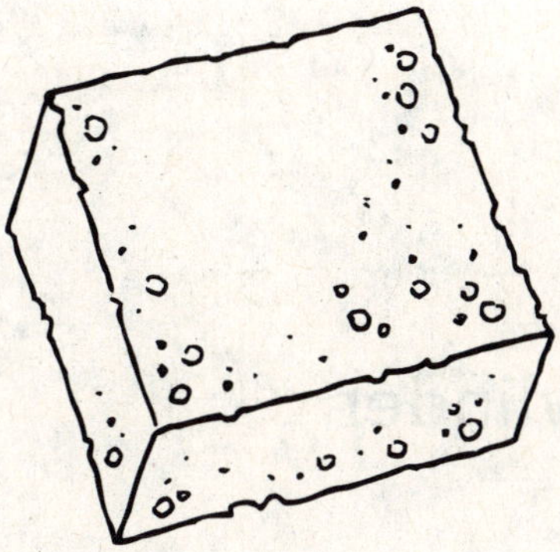

square rectangle

triangle trapezoid

2. A box of tissues has six square faces. What shape is the box of tissues? 1.G.1

____

3. Circle the shapes that are not used to make the solid. I.G.2

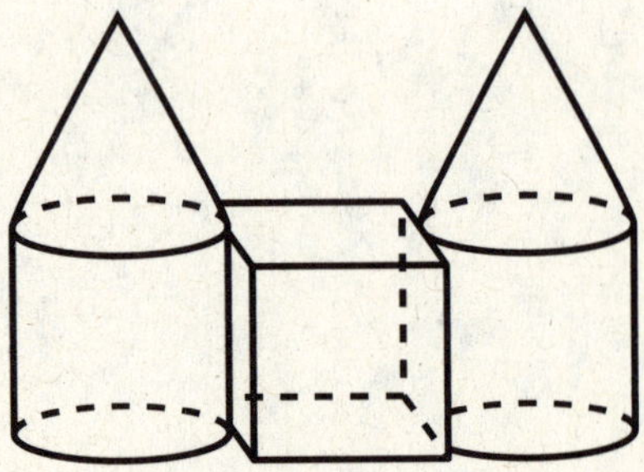

cone circle cylinder

cube triangle

4. Mr. Craft has a party hat that is shaped like a cone. How many vertices does his hat have? I.G.1

A. 0

B. 1

C. 2

5. Juanita made a pattern with blocks. What shape does she need to complete her pattern? I.G.I

cone

rectangular prism

cylinder

cube

6. Underline the shapes that have fewer than six vertices. I.G.I

cone

cylinder

cube

rectangular prism

7. Circle the shape best for packing books in. I.G.I

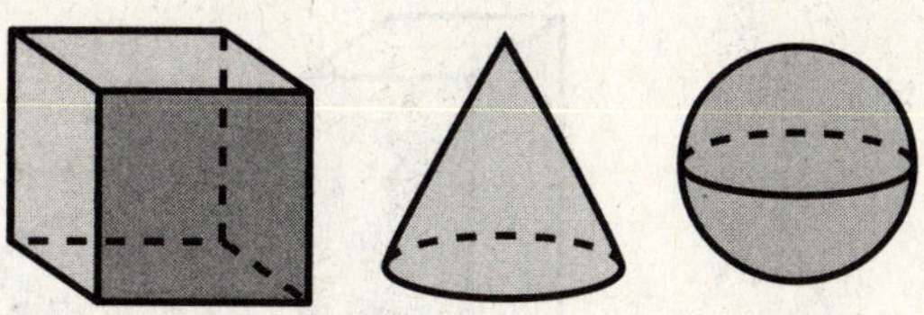

8. A can of tennis balls has 2 faces shaped like circles and no vertices. Circle the name of the shape. I.G.1

sphere　　cylinder　　cone

9. Mrs. Watson is thinking of a shape. All the faces of her shape are the same. What is Mrs. Watson's shape? I.G.1

A. cube

B. cone

C. cylinder

10. Circle the two shapes that have the same number of faces. I.G.1

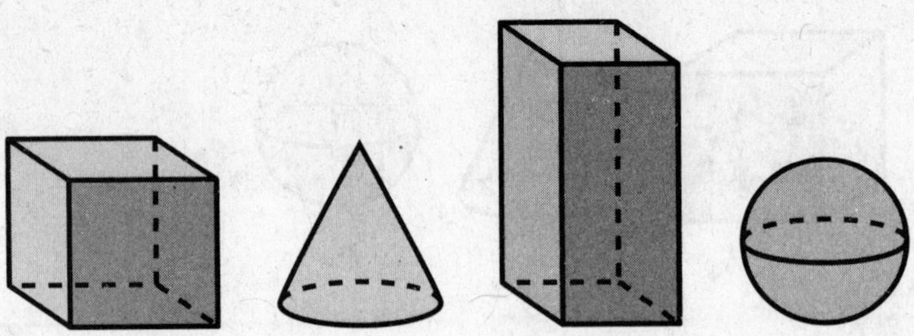

11. The two shapes are faces of a three-dimensional shape. What is the name of the three-dimensional shape? I.G.1

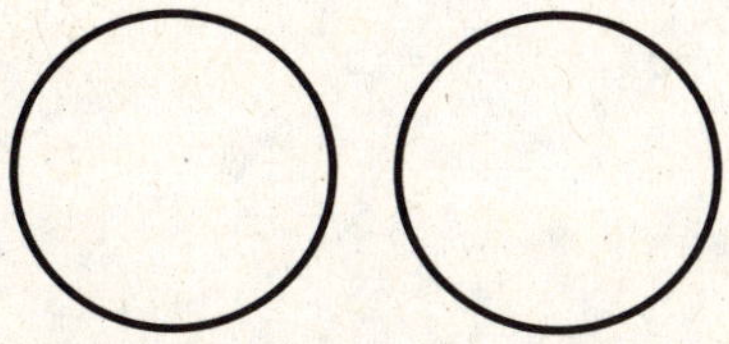

____

12. Sheldon has a collection of blocks that are cubes. He wants to store them in a container. What is the best shape for his container? I.G.2

A. rectangular prism

B. cone

C. cylinder

13. Which shape has a face that is the same shape as the bottom of a cylinder? I.G.I

A. cube

B. rectangular prism

C. cone

14. Which two shapes have the same number of vertices? I.G.I

- ☐ cone
- ☐ cylinder
- ☐ cube
- ☐ rectangular prism

15. Place an X on any shape that will not be in either of the blanks in the pattern. I.G.I

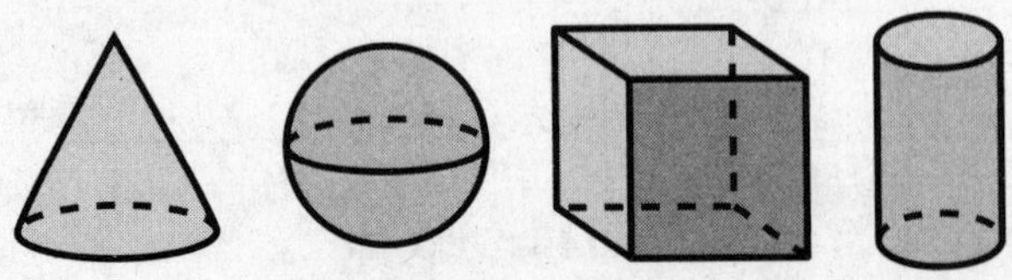

NAME

DATE

SCORE

Performance Task

A Lemonade Stand

Felipe is selling lemonade. He needs to buy lemons.

Write your answers on another piece of paper. Show all your work to receive full credit.

Part A

I.OA.I

Felipe fills two bags with lemons. Write an addition sentence to show the number of lemons in all.

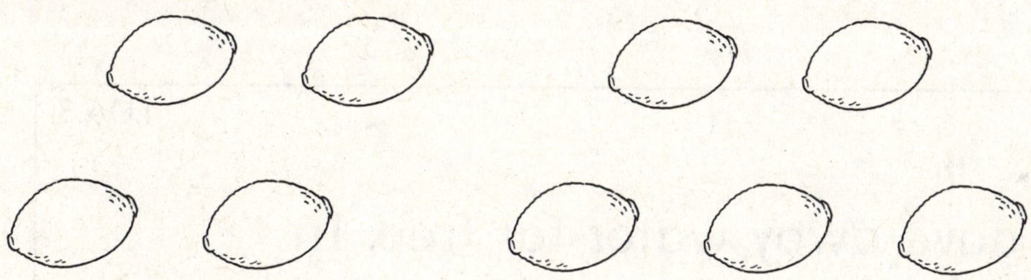

______ + ______ = ______ lemons

Part B

I.OA.I

Felipe sells 3 cups of lemonade in the morning and 5 cups in the afternoon. How many cups did Felipe sell in all?

______ + ______ = ______ cups

Performance Task *(continued)*

I.OA.6

Part C

10 people waved to Felipe. 7 were girls. Find the missing part to find out how many were boys.

Part	Part
7 girls	boys
Whole	
10 people	

I.OA.3

Part D

Felipe also gave away water for free. In the morning he gave away 7 cups of water. In the afternoon he gave away 0 cups of water. How many cups did he give away in all?

________ + ________ = ________ cups

NAME .... DATE ....

SCORE ....

Performance Task

At the Library

Mr. Fowler works at the library. He checks out books for people.

Part A

Mr. Fowler's first job of the day is checking in books that have been returned. Write a subtraction number sentence to show what he did.

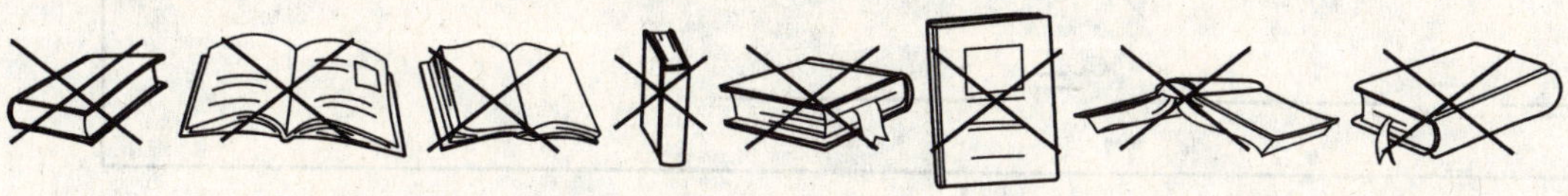

______ – ______ = ______

Part B

Mr. Fowler puts 8 books on his cart. He rolls the cart to the shelf and puts 5 books on it. How many books are left on the cart?

______ – ______ = ______

Performance Task *(continued)*

Part C

Mr. Fowler has a line of 5 people to help. He helps 3 people. How many people are left?

_______ – _______ = _______ people

Part D

Write the related subtraction sentence to show how many people Mr. Fowler helped.

_______ – _______ = _______

Part E

Mr. Fowler had 7 story time books on a shelf. The boys and girls who came to the library borrowed 3 books. Write the subtraction number sentence.

_______ – _______ = _______ books

NAME DATE

SCORE

Performance Task

Picking Apples

The Perez family is picking apples.

Write your answers on another piece of paper. Show all your work to receive full credit.

Part A

I.OA.6

Miguel picked 9 apples. His sister Juanita picked 7 apples. Use the number line to show how many apples they picked in all.

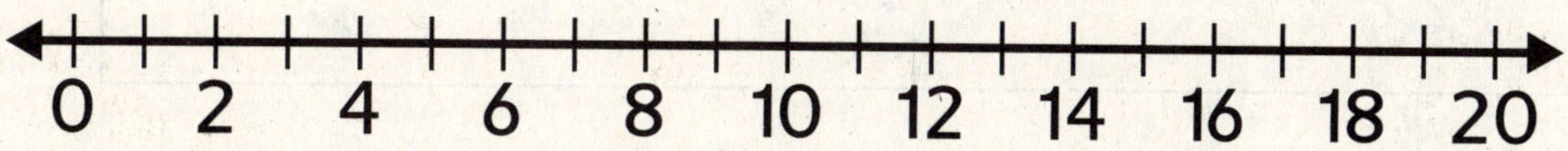

$$\begin{array}{r} 9 \\ +\ 7 \\ \hline \end{array}$$

apples

Performance Task *(continued)*

I.OA.3

Part B

Mrs. Perez picked 8 apples. Mr. Perez picked 4 apples. Write two addition sentences to find how many apples they picked in all.

_______ + _______ = _______ apples

_______ + _______ = _______ apples

I.OA.2

Part C

The other three children picked 5, 4, and 6 apples. Write a number sentence to find how many apples they picked in all.

_______ + _______ + _______ = _______ apples

NAME .... DATE ....

SCORE ....

Performance Task

Animals at the Zoo

Kerin takes a trip to the zoo with his class. He sees a lot of animals.

Write your answers on another piece of paper. Show all your work to receive full credit.

Part A

I.OA.5

Kerin counted 13 monkeys in a tree. 5 of the monkeys climb down. Use the number line to find how many monkeys are still in the tree.

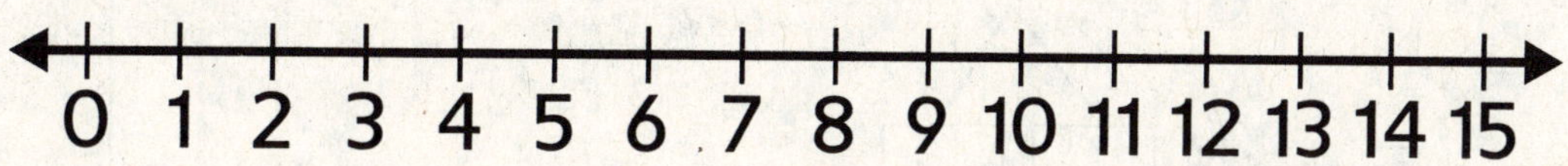

_______ − _______ = _______ monkeys left

Performance Task *(continued)*

I.OA.6

Part B

Write a related addition sentence for the monkeys Kerin saw.

_______ + _______ = _______ monkeys

I.OA.6

Part C

Kerin saw 11 ducks swimming in the pond. 4 ducks flew out of the water. How many ducks were left? Take apart the number to make 10. Then subtract.

11 − 4

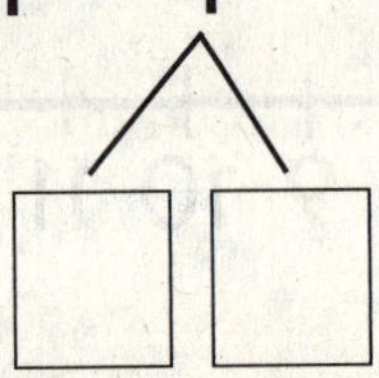

11 − 4 = _______ ducks left

I.OA.4

Part D

Kerin saw 15 lions. 5 were sleeping. How many lions were awake? Write a number sentence.

_______ − _______ = _______ lions

NAME ______ DATE ______

SCORE ______

Performance Task

Saving Money

Bree is saving money to buy a toy. She has nickels, pennies, and dimes.

Write your answers on another piece of paper. Show all your work to receive full credit.

Part A

I.OA.6

Bree has 8 pennies in her piggy bank. She finds 5 pennies on the ground. How much money does Bree have in all?

______ ¢ + ______ ¢ = ______ ¢

Performance Task *(continued)*

Part B

I.NBT.I

Bree takes the dimes out of her piggy bank. How much money does she have in dimes?

_____¢ _____¢ _____¢ _____¢ _____¢ _____¢ in all

Part C

I.NBT.I, I.NBT.3

Bree takes out her nickels. She has 85¢ in nickels. How many nickels does Bree have?

_________ nickels

Part D

I.NBT.3

Bree has 36 coins in all. Her brother Jack has 63 coins. Write < or >, and tell who has more coins.

36 ◯ 63

Who has more? _________

NAME ____________ DATE ____________

SCORE ____________

Performance Task

Buying School Supplies

Ms. Chatlos is buying supplies for her 1st grade class.

Write your answers on another piece of paper. Show all your work to receive full credit.

Part A

1.NBT.4

Ms. Chatlos buys 30 blue pens and 40 black pens. Write a number sentence to find how many pens she buys in all.

_______ + _______ = _______ pens

Part B

1.NBT.4

Ms. Chatlos buys 24 small notebooks and 30 big notebooks. Write a number sentence to find how many notebooks she buys in all.

_______ + _______ = _______ notebooks

Performance Task *(continued)*

I.NBT.6

Part C

Ms. Chatlos buys 40 boxes of crayons. She returns 20 boxes. How many boxes does she have left? Use the number line.

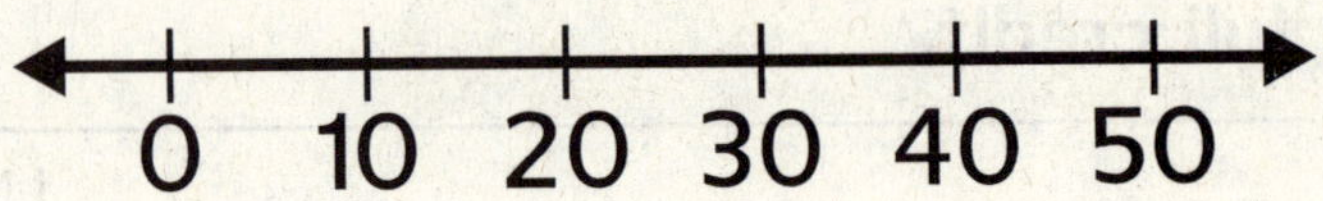

________ − ________ = ________ boxes

I.NBT.4

Part D

Ms. Chatlos has 8 pencils. She buys 31 more. How many pencils does she have in all? Write a number sentence.

________ + ________ = ________ pencils

NAME DATE

SCORE

Performance Task

Favorite Sport

Jude asked the students in his class their favorite sport. The table shows how many students chose each sport.

Favorite Sport	
Soccer	11
Football	2
Baseball	6

Write your answers on another piece of paper. Show all your work to receive full credit.

Part A

1.MD.4

Write the tallies for each sport.

Favorite Sports		
Sport	**Tally**	**Total**
Soccer		11
Football		2
Baseball		6

Performance Task *(continued)*

I.MD.4

Part B

What sport do people like the least?

Answer: ____________

I.MD.4

Part C

Make a bar graph.

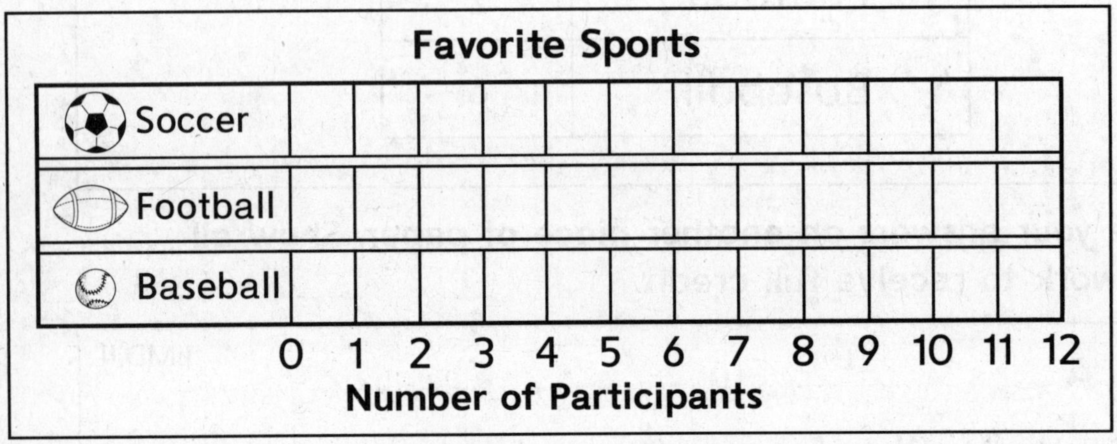

I.MD.4, I.NBT.4

Part D

How many students are in Jude's class?
Write a number sentence.

NAME ______ DATE ______

SCORE ______

Performance Task

Life on a Farm

A farmer is growing corn. He needs to measure how tall it grows.

Write your answers on another piece of paper. Show all your work to receive full credit.

Part A

I.MD.3

The farmer goes to the field at 6:00. It takes two hours to measure the corn. What time does he finish? Show the time on the clock.

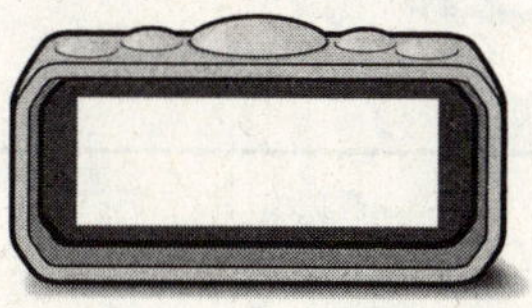

Part B

I.MD.3

The farmer finishes eating lunch at 1:00. He spent one hour eating. What time did he start eating lunch? Draw the time on the clock.

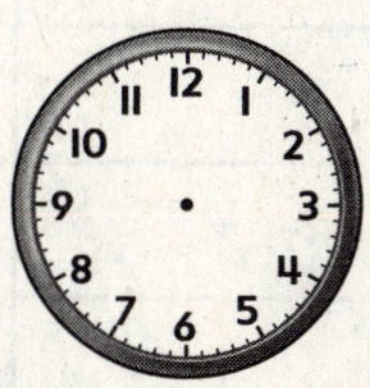

Performance Task *(continued)*

Part C I.MD.3

The farmer picks some corn for dinner. He starts picking corn at the time shown on the clock.

He spends a half an hour picking corn. What time does he finish picking corn?

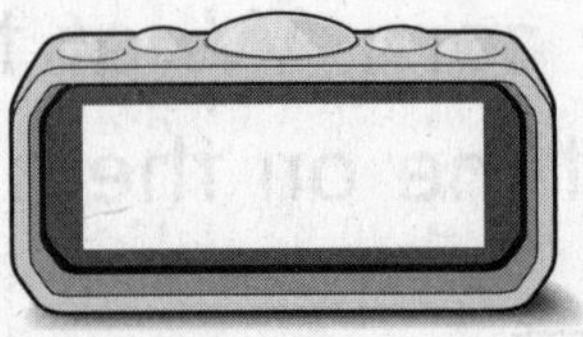

Part D I.MD.2

On Monday the corn was 5 cubes high. It grows 4 cubes higher every day. How many cubes high is it when the farmer measures on Thursday?

Days	Cubes
Monday	5
Tuesday	
Wednesday	
Thursday	

NAME ________ DATE ________

SCORE ________

Performance Task

Building a Playground

A school wants to build a playground. The shape will be a heptagon. Research what a heptagon looks like.

Write your answers on another piece of paper. Show all your work to receive full credit.

Part A

I.G.I

A heptagon has 7 sides. Draw a heptagon. How many vertices does it have?

________ vertices

Performance Task *(continued)*

Part B I.G.I

Part of the playground will have slides. This part will be in this shape.

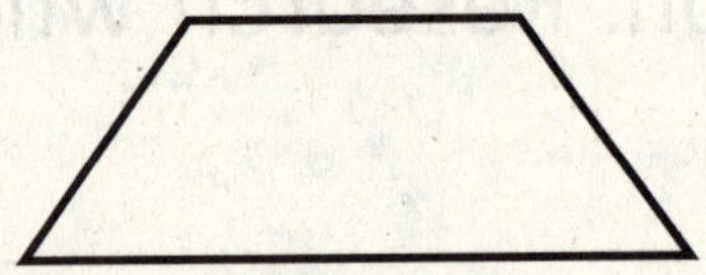

What is this shape called? ______

Part C I.G.I, I.G.2

Chunks of grass come in triangle and rectangle shapes. Draw two lines to make a rectangle and two triangles.

Part D I.G.I, I.G.3

The swing area is a rectangle. It needs divided into fourths. Draw the swing area. Show it divided into fourths.

NAME DATE

SCORE

Performance Task

Making a Castle Wall

Jerome is building a castle wall out of blocks.

Write your answers on another piece of paper. Show all your work to receive full credit.

Part A

I.G.2

The first part of the wall looks like this. Circle the names of the shapes that are used to make the wall.

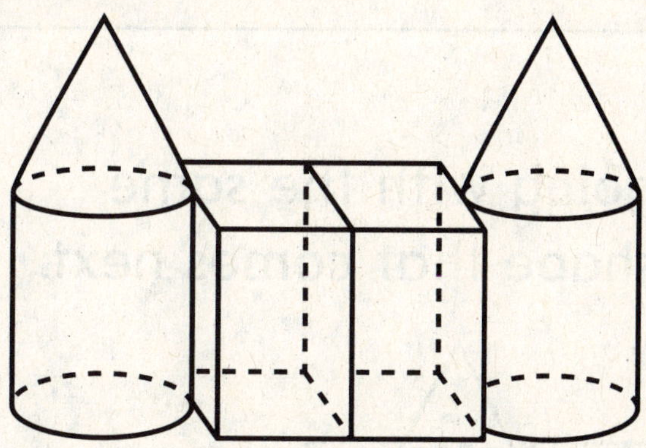

cylinder cube

rectangular prism cone

Performance Task *(continued)*

I.G.1

Part B

How many vertices does each shape have?

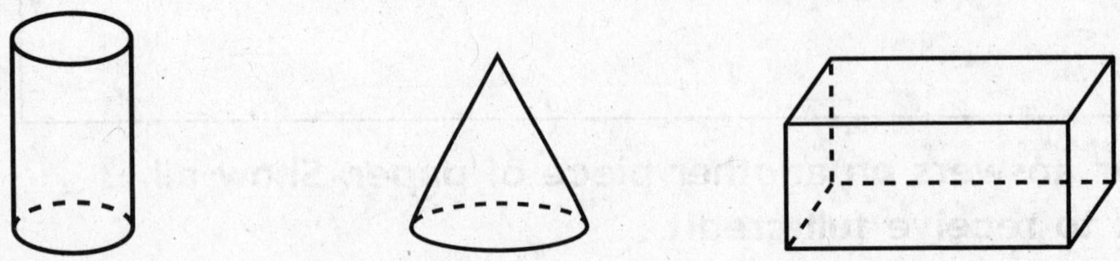

__________ __________ __________

Which one has the least? ________________

I.G.1

Part C

The wall will keep going with the same pattern. Circle the shape that comes next.

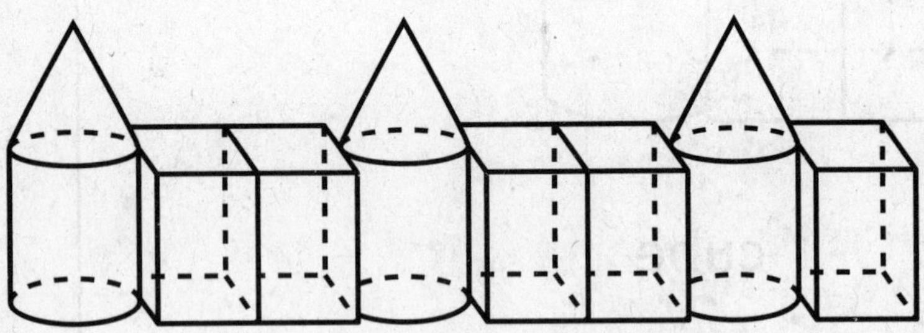

____

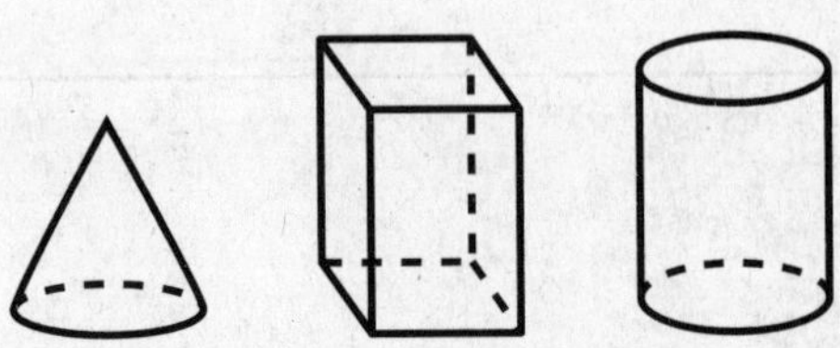

NAME .... DATE ....

Benchmark Test I

SCORE ....

1. Maddie has 4 pennies in her pocket. She has 8 pennies on her desk. How many pennies does Maddie have in all? I.OA.6

4¢ + 8¢ = ______ ¢

2. Add zero. I.OA.3

7 + ______ = ______

3. True or false? I.OA.I, I.OA.7

Tyler has 8 cars. He gives I to his brother. Now Tyler has 7 cars.

True False

4. Use the number line to solve. Write the sum. 1.OA.6

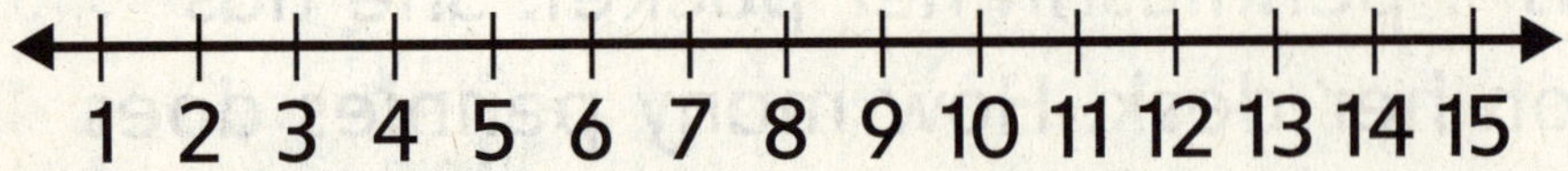

$$\begin{array}{r} 7 \\ +\ 8 \\ \hline \end{array}$$

5. True or false?

Zachary has 7 blocks. He picks up 1 more block. Zachary has 8 blocks in all. 1.OA.1, 1.OA.7

True   False

6. Subtract. 1.OA.6

$$\begin{array}{r} 10 \\ -\ 6 \\ \hline \end{array}$$

7. Find the matching addition number sentence. I.OA.I

A. $3 + 4 = 6$

B. $3 + 4 = 7$

C. $2 + 4 = 6$

8. Circle the correct doubles facts. I.OA.6

$6 + 6 = 11$ $8 + 8 = 16$

$4 + 4 = 6$ $9 + 9 = 18$

9. Write the related subtraction fact. I.OA.6

$7 - 6 = 1$

______ − ______ = ______

10. Abigail's teacher asked her to add 7 + 5. She makes 10 to add. Which number sentence does Abigail use? 1.OA.6

10 + 1 = 11

10 + 2 = 12

10 + 3 = 13

11. Find the missing part. 1.OA.1

Part	Part
2	
Whole	
7	

12. Write a number sentence. 1.OA.1

There were 6 pieces of pizza. Andre eats some of the pieces. There are 3 left. How many did Andre eat?

_____ ◯ _____ = _____ pieces

13. True or false? Mr. Valencia has 8 red ties and 6 blue ties. He has 8 blue socks and 6 black socks. Mr. Valencia has the same number of ties as socks. 1.OA.3, 1.OA.7

True　　　　　　False

14. Geraldine has 8 pencils. Circle the ways that will make 8 pencils. 1.OA.6

$2 + 6$　　　　　　$4 + 3$

$5 + 2$　　　　　　$0 + 8$

15. Mrs. Clark buys 5 red balloons, 7 orange balloons, and 3 yellow balloons. How many balloons did she buy in all? 1.OA.2

______ + ______ + ______ = ______ balloons

16. Find the matching subtraction number sentence. I.OA.I, I.OA.5

A. $8 - 5 = 3$

B. $7 - 5 = 3$

C. $5 - 8 = 3$

17. Antwan read 7 books in January. He read 9 books in February. How many books did he read in all? I.OA.6

_______ books

18. Circle two numbers that will make 10. I.OA.6

2	6	3
7	1	5

19. Add. 1.OA.6

$$\begin{array}{r} 7 \\ +\ 7 \\ \hline \end{array}$$

20. Find the related subtraction sentence. 1.OA.6

$$10 - 4 = 6$$

A. $6 - 4 = 2$

B. $10 - 2 = 8$

C. $10 - 6 = 4$

Performance Task

Collecting Cans

Paul is collecting cans to recycle. He picks up a half-dozen cans. Research how many cans are in a half-dozen.

Write your answers on another piece of paper. Show all your work to receive full credit.

Part A

1.OA.1

On Monday, Paul starts with a half-dozen cans. He finds 3 more. How many cans does Paul have? Write a number sentence.

Part B

1.OA.1

On Tuesday, Paul collects 7 cans. He carries 4 in his right hand and the rest in his left hand. How many cans are in his left hand? Write a number sentence.

Part C

1.OA.6

Write the related subtraction sentence for the cans he collects on Tuesday.

NAME ________ DATE ________

Benchmark Test 2

SCORE ________

1. Circle the correct answer. I.NBT.4

$$\begin{array}{r} 30 \\ +\ 30 \\ \hline \end{array}$$

A. 6

B. 16

C. 60

2. Carlina has the boxes of tissues shown. Circle a group of ten boxes. Write how many more. How many boxes does Carlina have in all? I.NBT.2b

10 and ______ more ______ boxes

3. True or false? 4 tens and 7 ones is the same as forty-seven. I.NBT.5

True False

4. 20 + ______ = 90. Mark the related subtraction fact. I.NBT.6

☐ 20 + 60 = 90 ☐ 90 − 20 = 70

☐ 90 − 20 = 60 ☐ 70 + 20 = 90

5. Ted starts counting backwards from 17. He says, "17, 16, 15, 14, 13, 12." Circle the number sentence is he trying to solve. I.OA.5

17 − 5 = 12 17 − 4 = 12

17 − 4 = 13 17 − 3 = 13

6. Are the number sentences part of the fact family? Circle yes or no. I.NBT.6

yes no 40 + 40 = 80

yes no 80 − 40 = 40

yes no 40 − 80 = 40

7. Damian has 53 stamps. He buys 30 more stamps. How many stamps does Damian have in all? I.NBT.4

$$\begin{array}{r} 53 \\ +\ 30 \\ \hline \end{array}$$

_______ stamps

8. Barret has 62 ribbons. How many groups of ten does he have? How many ones? I.NBT.2a

_______ tens and _______ ones

9. There are 16 blocks. Katherine loses 5 of them. How many blocks does she have left? Use the number line to help you subtract. I.OA.5

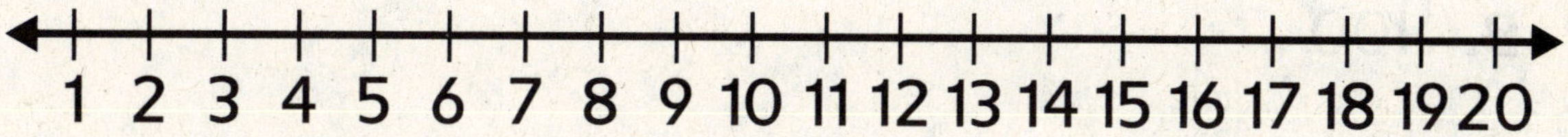

_______ blocks

10. Use the number line to subtract. Write the difference. I.NBT.6

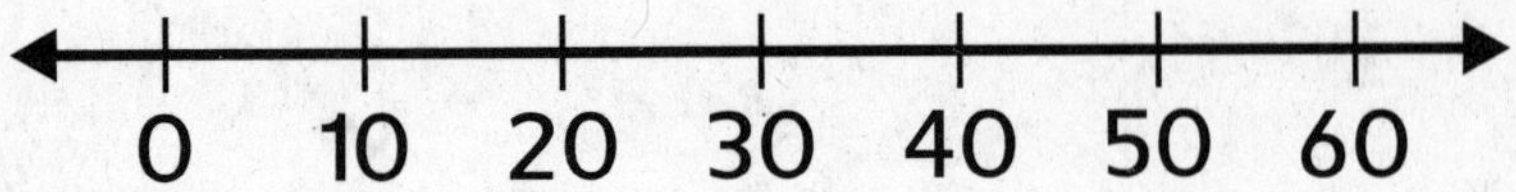

50 – 40 = ______

11. True or false? Deangelo has 43 marbles. Kumar has 63 marbles. Kumar has 20 more marbles than Deangelo. I.NBT.5

True False

12. Elise is thinking of a number between 99 and 101. What is her number? I.NBT.1

A. 98

B. 100

C. 102

13. Manuel saw two kinds of animals at the zoo. He saw 14 animals in all. Circle the kinds of animals Manuel saw. I.OA.3, I.OA.7

6 Leopards

8 Lions

4 Tigers

14. Mr. Huan drinks 10 cups of water every day. How many cups does he drink in 5 days? Fill in the table. I.NBT.2c

Days	Cups

15. Miranda buys a greeting card that costs 55¢. She only used nickels to pay. How many nickels did Miranda use? 1.NBT.1

_______ nickels

16. Luciana has 13 granola bars. She gives away 5 of them. How many granola bars does she still have? Write an addition sentences and a subtraction sentence. 1.OA.6

_______ + _______ = _______

_______ − _______ = _______ granola bars

17. Gianna is thinking of a number. When she adds 3 to the number, she gets 10. What is her number? 1.OA.4

3 + _______ = 10

18. Mrs. Klein has 16 pencils. She buys 8 more. How many pencils does she have in all? Circle the ones to show regrouping. Write your answer. 1.NBT.4

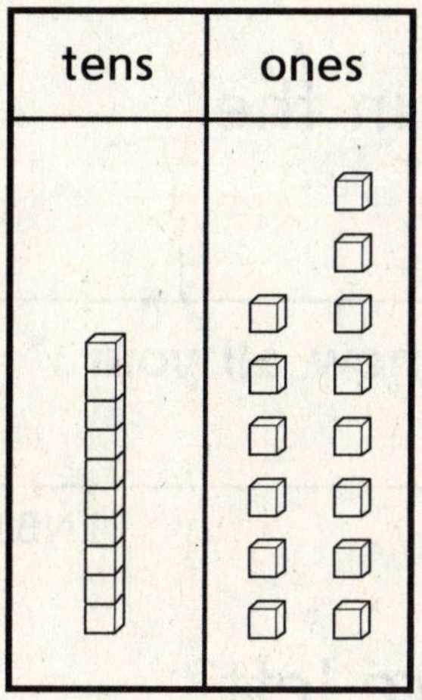

________ pencils

19. Which subtraction problem has the same answer as 16 − 7? 1.OA.6

10 − 1

10 − 2

10 − 3

20. Mia has 4 red cups and some orange cups. She has 14 cups in all. How many orange cups does Mia have? 1.OA.4

________ orange cups

Performance Task

United States Senators

Research how many senators there are in the United States Senate.

Write your answers on another piece of paper. Show all your work to receive full credit.

Part A

I.NBT.6

If 30 senators are absent, how many are left? Write a number sentence and use the number line to answer.

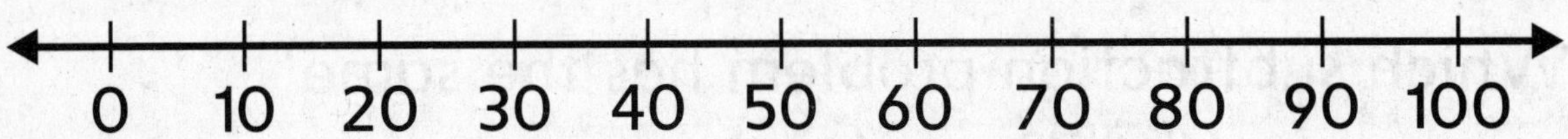

Part B

I.NBT.4

25 senators were in a meeting. 3 more senators showed up. How many senators are there in the meeting? Write a number sentence.

Part C

I.NBT.4

16 senators made a speech on Monday. 20 senators made a speech on Tuesday. How many senators spoke in all? Write a number sentence.

NAME DATE

Benchmark Test 3

SCORE

1. Olivia grew a green bean plant. It had 9 beans on it. Olivia ate 5 green beans. How many green beans are left? 1.OA.1

   A. 9 − 5 = 3 green beans

   B. 9 − 5 = 4 green beans

   C. 9 − 4 = 5 green beans

2. Find the missing part of 10. 1.OA.4, 1.OA.8

Part	Part
6	
Whole	
10	

3. Write the addition number sentence. 1.OA.1, 1.OA.1

______ + ______ = ______

4. Write the related subtraction number sentence. 1.OA.6

$$4 - 3 = 1$$

______ − ______ = ______

Subtract. 1.OA.6

5. $6 - 2 =$ ______

6. $10 - 3 =$ ______

7. True or false? 1.OA.1, 1.OA.7

Jamal has 8 apples. He eats 3. Now Jamal has 5 apples.

True False

8. Make 10 to add. 1.OA.3, 1.OA.6

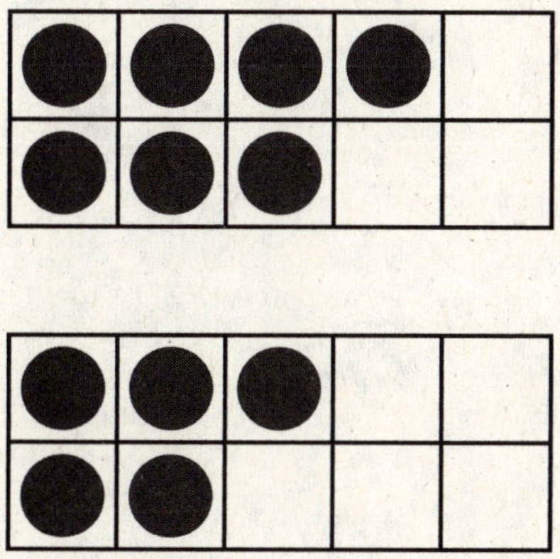

$$\begin{array}{r} 7 \\ +\ 5 \\ \hline \end{array} \longrightarrow \begin{array}{r} 10 \\ +\ \\ \hline \end{array}$$

9. Circle the greater number. Count on to add. 1.OA.5, 1.OA.6

3 + 9 = ______

Find the sum. I.OA.6

10. $\begin{array}{r} 8 \\ +\ 5 \\ \hline \end{array}$

11. $\begin{array}{r} 6 \\ +\ 9 \\ \hline \end{array}$

12. 14 − 6 I.OA.6

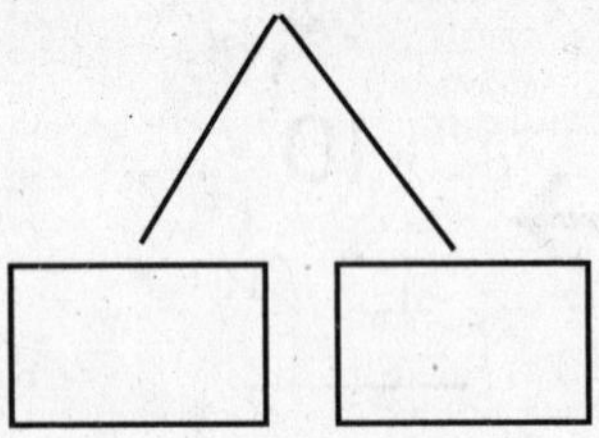

________ − ________ = ________

________ − ________ = ________

So, 14 − 6 = ________

13. 18 − 9 = ________ I.OA.6

14. Complete the fact family. 1.OA.3, 1.OA.6

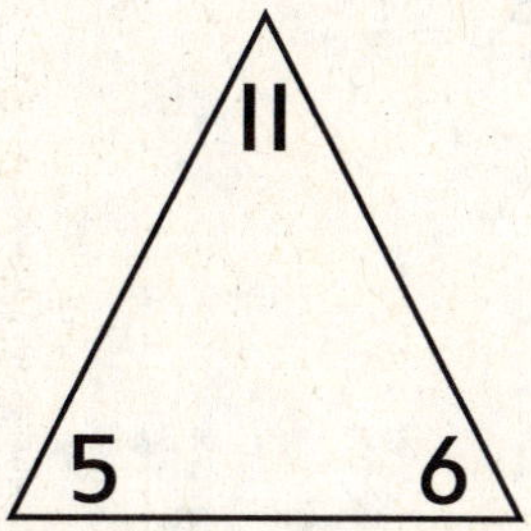

5 + 6 = ______ 11 − 6 = ______

6 + 5 = ______ 11 − 5 = ______

15. Complete the place value chart. 1.NBT.2

tens	hundreds	ones

2	7	8

16. Compare. Use >, <, or =. 1.NBT.3

34 ◯ 28

17. Josiah bakes 4 groups of ten rolls. He bakes 3 more rolls. How many rolls did he bake in all? 1.NBT.4

____

18. Read the numbers. What number is missing? 1.NBT.1

117, ________, 119, 120

A. 121

B. 115

C. 118

19. Subtract. 1.NBT.2c, 1.NBT.6

7 tens − 5 tens = ________ tens

________ − ________ = ________

20. Add. I.NBT.4

	tens	ones
	2	2
+	0	4

21. Count on to add. I.NBT.4

86 + 3 = ______

22. Complete the tally chart. I.MD.4

Favorite Pet		
Type	Tally	Total
Cat	卌 |||	
Dog	卌		
Fish	||||	

23. Use the tally chart in Exercise 22 to make a bar graph. I.MD.4

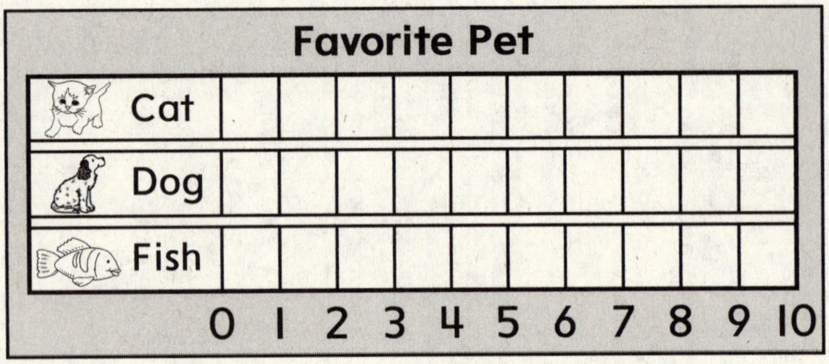

24. Use the graphs above to answer the question. How many people chose fish and cats? I.MD.4

____

25. Order the objects by length. Write 1 for long, 2 for longer, and 3 for longest. I.MD.1

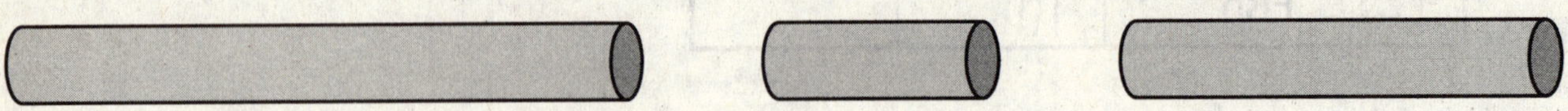

________ ________ ________

26. Write the time on the digital clock. I.MD.3

half past 4

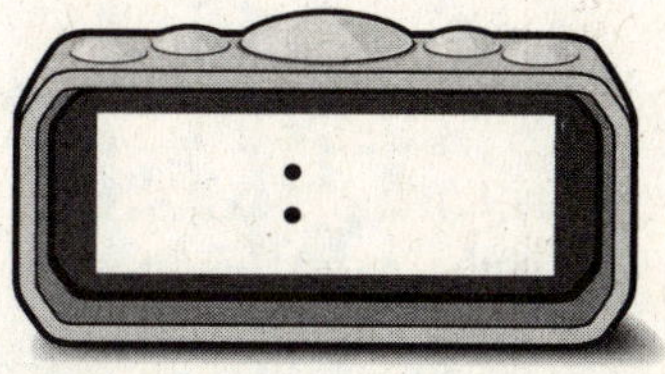

27. Write the time on the clock. I.MD.3

Piper gets up at 7:30. Her bus comes 1 hour later. What time does the bus come?

28. Write how many sides and vertices. I.G.1

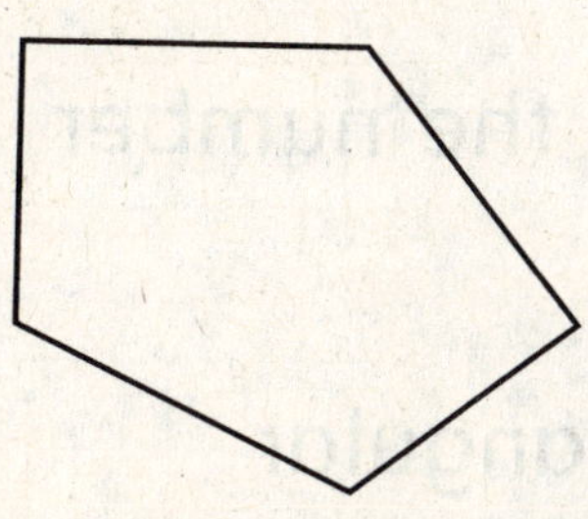

____

____

29. Write how many parts are shaded. I.G.3

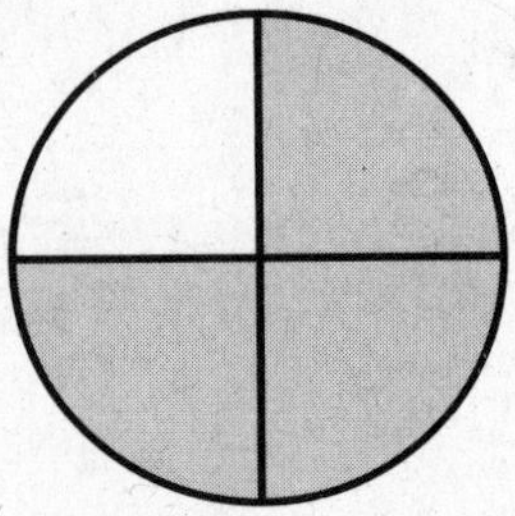

________ of ________ parts

30. Circle the shape that is not used to build the composite shape below. I.G.2

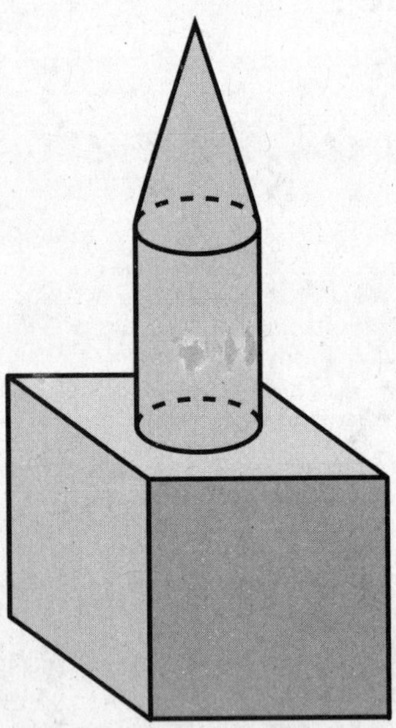

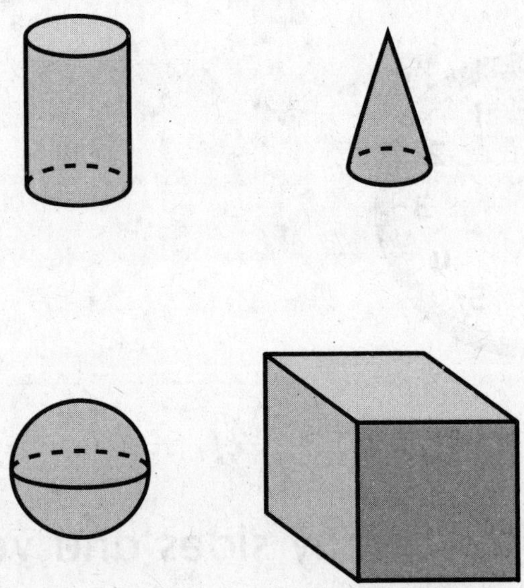

31. Circle the name of the shape. Write the number of faces and vertices. I.G.1

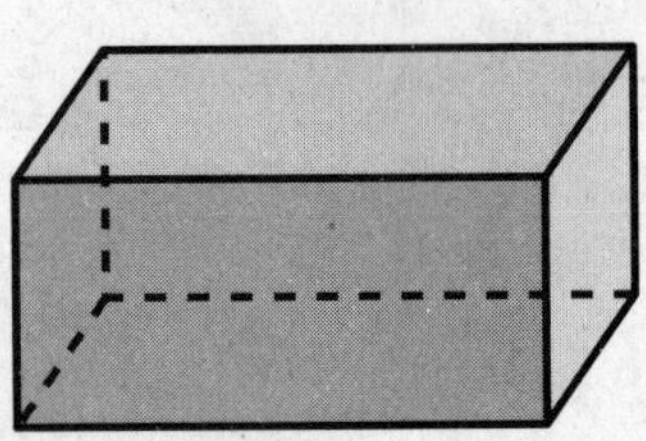

cone

rectangular prism

____ faces

____ vertices

Performance Task

Building Blocks

Liam and Genji are playing with wooden blocks.

Part A

Liam compares a cube and a cylinder. Which one has more faces? Write the number of faces below each block. Circle the one with more faces.

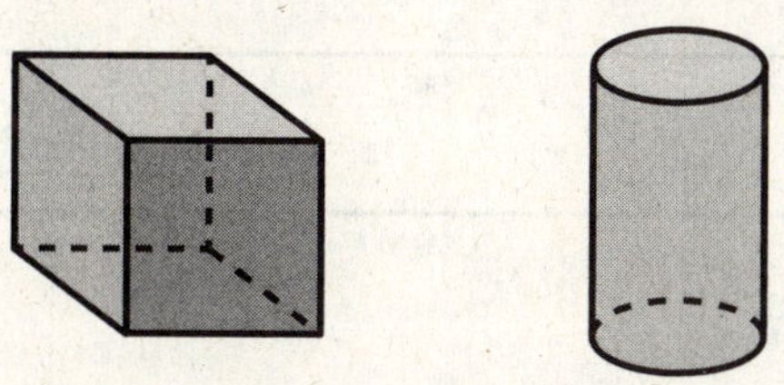

______ ______

Part B

Genji compares a cube and a cone. Which one has fewer vertices? Write the number of vertices below each block. Circle the block with fewer vertices.

______ ______

Performance Task *(continued)*

Part C

Genji and Liam make a pattern with the blocks. Circle the missing shape.

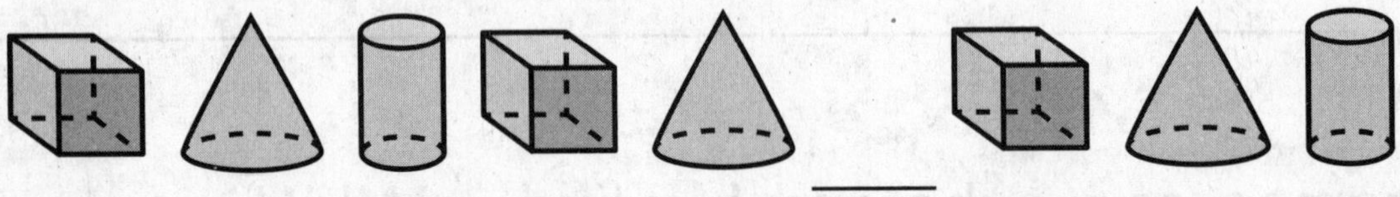

Part D

Genji and Liam build a tower. Draw lines connecting the blocks to where they fit in the tower.

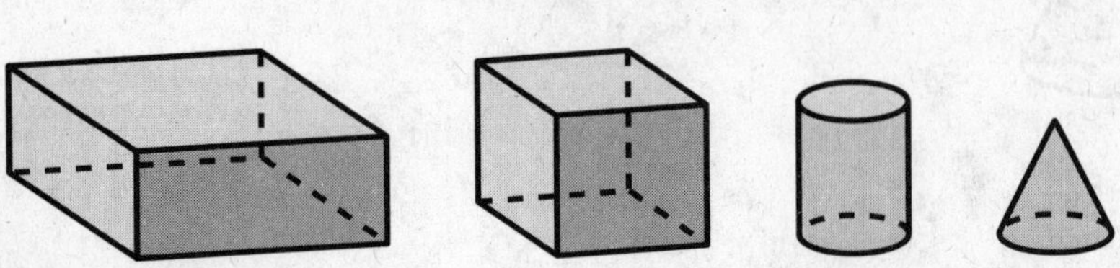

NAME .... DATE ....

Benchmark Test 4

SCORE ....

1. Write the addition number sentence. I.OA.I

_____ + _____ = _____

2. 12 − 5 = _____ I.OA.6

3. Write the time on the digital clock. I.MD.3

half past two

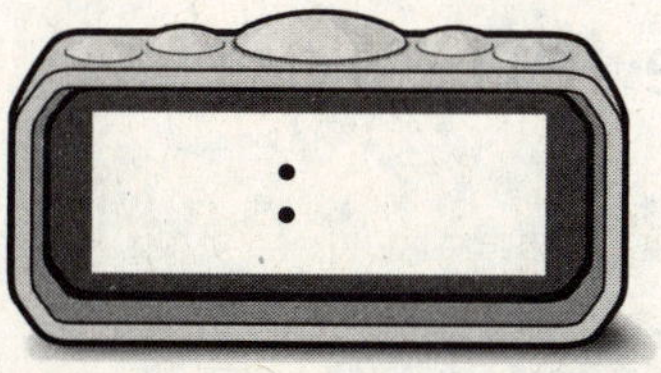

4. 17 − 9 1.OA.6

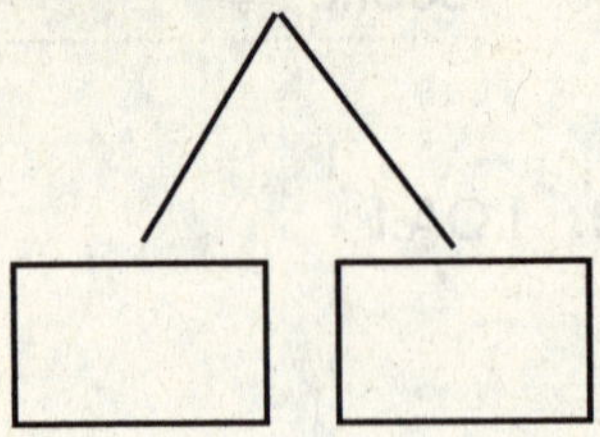

________ − ________ = ________

________ − ________ = ________

So, 17 − 9 = ________.

5. Count on to add. 1.NBT.4

72 + 5 = ________

6. Janice had a pizza with 8 pieces in it. Janice ate 3 pieces. How many pieces are left? 1.OA.1

A. 8 − 3 = 5 pieces

B. 8 − 5 = 4 pieces

C. 8 − 5 = 3 pieces

7. Complete the place value chart. I.NBT.2

tens	hundreds	ones

1	6	4

8. Write the related subtraction number sentence. I.OA.6

$6 - 2 = 4$

________ − ________ = ________

9. Write the time on the clock. I.MD.3

Veronica starts reading at 6:30. She goes to bed one hour later. What time does she go to bed?

10. Write how many sides and vertices. I.G.1

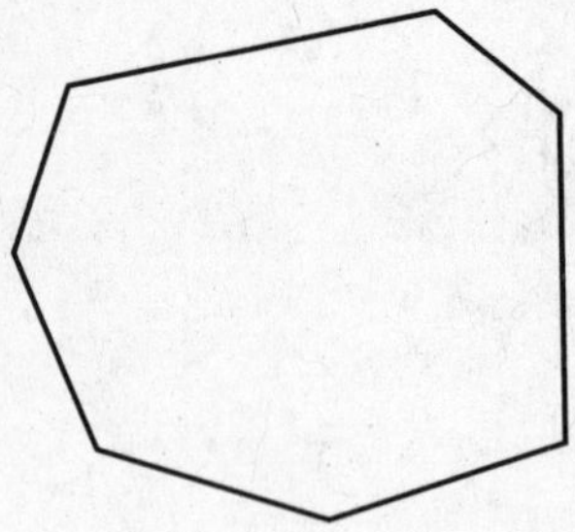

_______ sides

_______ vertices

11. Complete the fact family. I.OA.3, I.OA.6

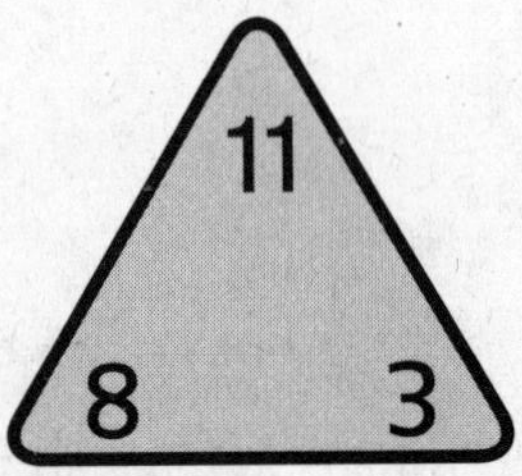

8 + 3 = _______

3 + 8 = _______

11 − 8 = _______

11 − 3 = _______

12. Find the missing part of 10. I.OA.4, I.OA.8

Part	Part
7	
Whole	
10	

13. Complete the tally chart. I.MD.4

Favorite Season		
Type	**Tally**	**Total**
Summer		9
Fall		5
Winter		6
Spring		2

14. Use the tally chart in Exercise 13 to make a bar graph. I.MD.4

Favorite Season

Summer										
Fall										
Winter										
Spring										
	0	1	2	3	4	5	6	7	8	9	10

15. Use the graphs above to answer the question. How many people chose fall or winter? I.MD.4

____

Subtract. I.OA.6

16. 7 − 5 = ________

17. 10 − 5 = ________

18. Add. I.NBT.4

	tens	ones
	3	1
+		5

19. Compare. Use >, <, or =. I.NBT.3

28 ◯ 82

20. True or false? I.OA.I, I.OA.7

Brayden had 9 books. He took 4 back to the library. Now Brayden has 5 books.

True   False

21. Write how many parts are shaded. I.G.3

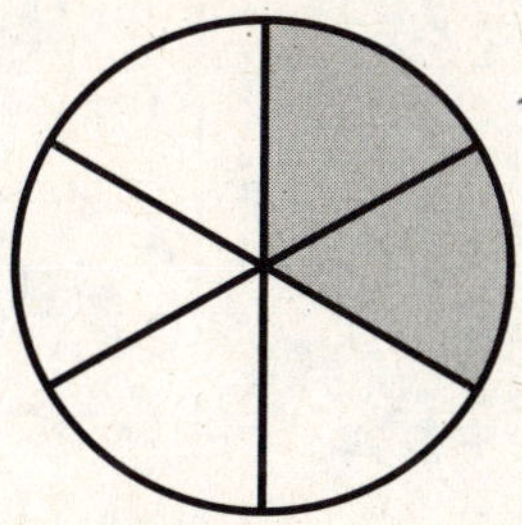

________ of ________ parts

22. Marcus buys 4 groups of ten baseball cards. He buys 2 more cards. How many cards did he buy in all? I.NBT.4

________ cards

23. Circle the name. Write the number of faces and vertices. 1.G.1

cone　　　　cube

cylinder　　　　rectangular prism

________ faces　　________ vertices

24. Make 10 to add. 1.OA.3, 1.OA.6

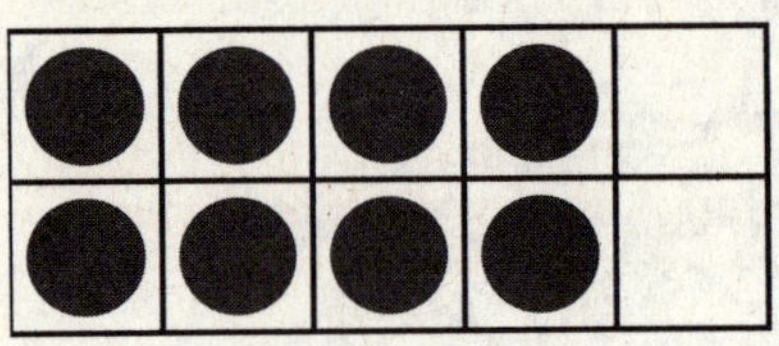

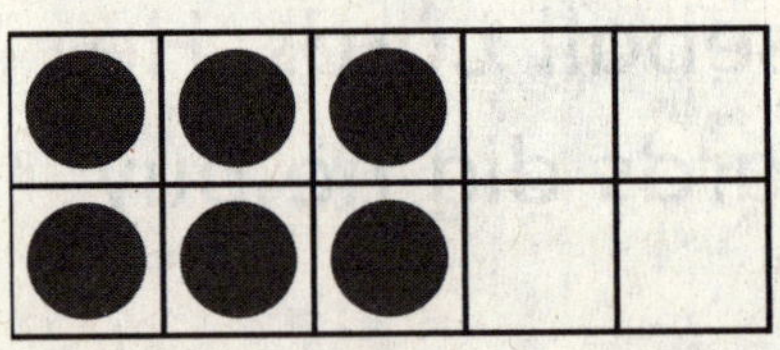

$$\begin{array}{r} 8 \\ +\ 6 \\ \hline \end{array} \longrightarrow \begin{array}{r} 10 \\ +\ \\ \hline \end{array}$$

25. Find the sum. 1.OA.6

$$\begin{array}{r} 7 \\ +\ 4 \\ \hline \end{array}$$

26. Find the sum. I.OA.6

$$\begin{array}{r} 6 \\ +\ 8 \\ \hline \end{array}$$

27. Circle the shape that is not used to build the composite shape below. I.G.2

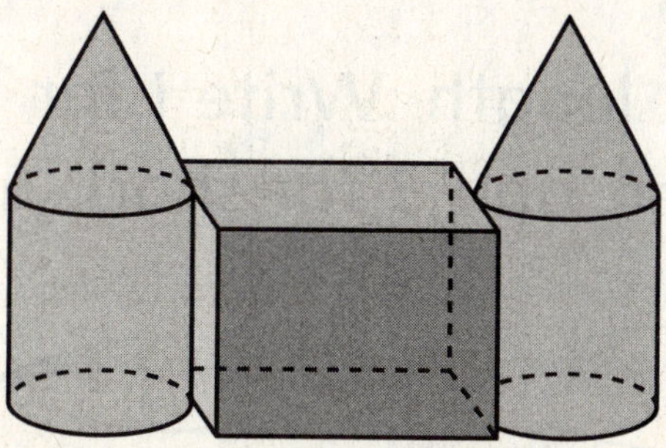

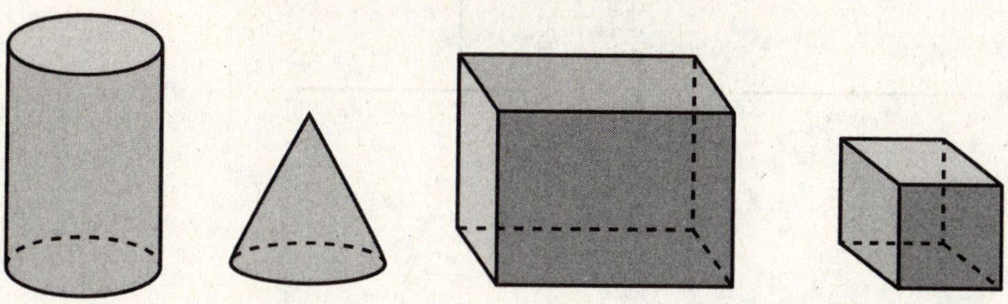

28. Subtract. I.NBT.2c, I.NBT.6

8 tens − 3 tens = ________ tens

80 − ________ = ________

29. Circle the greater number. Count on to add. I.OA.5, I.OA.6

5 + 8 = ________

30. Read the numbers. What number is missing? I.NBT.I

I08, I09, _____, III

A. I00

B. I0I

C. II0

31. Order the objects by length. Write I for shortest, 2 for shorter, and 3 for short. I.MD.I

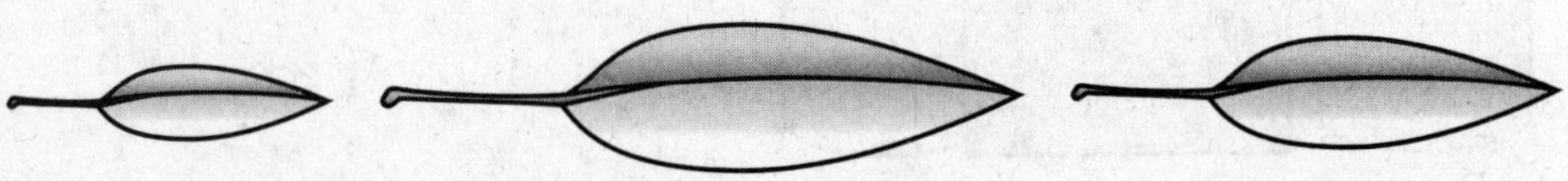

_____ _____ _____

Performance Task

Donating Books for a Library

Mrs. Boylan's class is collecting books to donate to a library.

Write your answers on another piece of paper. Show all your work to receive full credit.

Part A

1.NBT.3

Kim collected 23 books. Ranata collect 31 books. Deshawn collect 46 books.

Who collected the most? ____________

Who collected the least? ____________

Part B

1.NBT.4

On the first day, the class collected 36 books. On the second day, they collected 20 books. How many books were collected in all? Write a number sentence.

Performance Task *(continued)*

I.OA.5

Part C

Henry collects 37 books. 5 of them were picture books. How many were not picture books? Use the number line to count backwards.

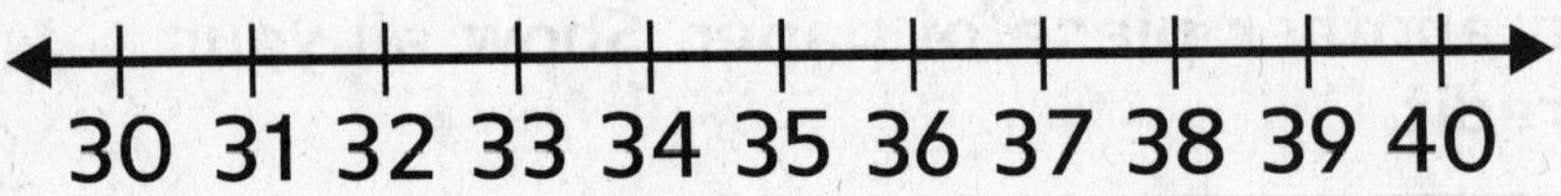

_______ books

I.G.I

Part D

The class packs the books in boxes like this. How many faces and vertices do the boxes have?

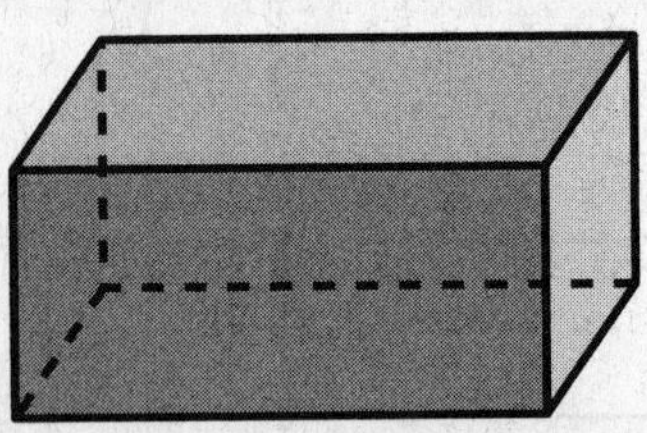

_______ faces

_______ vertices

I.MD.3

Part E

The class carries the books to the library. They start carrying the books at half past 10. It takes one hour to carry the books. Draw the time they finish.

Smarter Balanced Assessment Item Types

In Grade 3, students will begin to take the Smarter Balanced Test for Math. It is a test that is taken on a computer. The problems on the next few pages show the kinds of questions students might have to answer and what to do to show the answer on the computer. However, the student materials in the Grade 1 book center mainly on ***Multiple Choice, Drag and Drop,*** and ***Click to Select*** formats.

Selected Response means that you are given answers from which you can choose.

Selected Response Items

Regular multiple choice questions are like tests you may have taken before. Read the question and then choose the one best answer.

Multiple Choice

Sasha has a collection of coins. There are 3 in one pile and 2 in another. How many coins does she have in all?

☐ 2

☐ 3

■ 5

ONLINE EXPERIENCE Click on the box to select the one correct answer.

HELPFUL HINT Only one answer is correct. You may be able to rule out some of the answer choices because they are unreasonable.

Sometimes a multiple choice question may have more than one answer that is correct. The question may or may not tell you how many to choose.

Multiple Correct Answer

A fact family is made from the numbers 3, 7, and 10. Choose **two** statements that belong to this fact family.

■ $3 + 7 = 10$

■ $10 - 3 = 7$

☐ $5 + 5 = 10$

☐ $10 - 6 = 4$

ONLINE EXPERIENCE Click on the box to select it.

HELPFUL HINT Read each answer choice carefully. There may be more than one right answer.

Another type of question asks you to tell whether the sentence given is true or false. It may also ask you whether you agree with the statement, or if it is true. Then you select yes or no to tell whether you agree.

Multiple True/False or Multiple Yes/No

There were no birds on the pond. Two brown ducks swim into the pond. Five geese fly in and land on the pond. Tell whether you agree with each sentence.

Yes	No	
■	☐	There are more geese than ducks on the pond.
☐	■	There are 2 birds on the pond.
☐	■	5 geese flew away.
■	☐	There are 7 birds in all on the pond.

ONLINE EXPERIENCE Click on the box to select it.

HELPFUL HINT There is more than one statement. Any or all of them may be correct.

You may have to choose your answer from a group of objects.

Click to Select

Which model shows $4 - 1 = 3$?

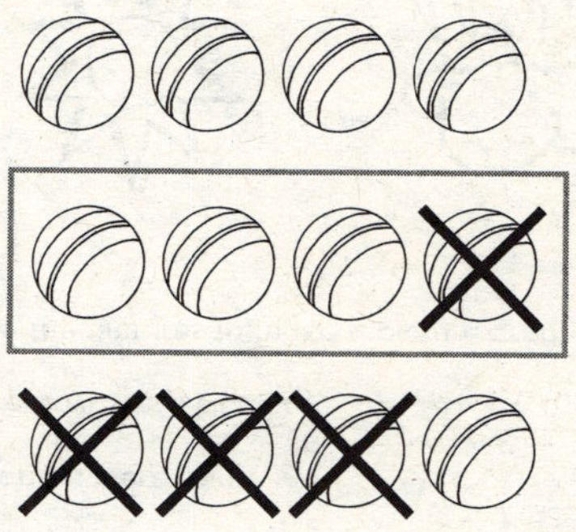

ONLINE EXPERIENCE Click on the figure to select it.

HELPFUL HINT On this page you can draw a circle or a box around the figure you want to choose.

When no choices are given from which you can choose, you must create the correct answer. One way is to type in the correct answer. Another may be to make the correct answer from parts that are given to you.

Constructed-Response Items

Fill in the Blank

What time is shown on the clock?

Write the time on the digital clock.

ONLINE EXPERIENCE You will click on the space and a keyboard will appear for you to use to write the numbers and symbols you need.

HELPFUL HINT You must write all of the digits for the time. For example, do not forget to include the minutes if the time shown is "6 o'clock."

Sometimes you must use your mouse to click on an object and drag it to the correct place to create your answer.

Drag and Drop

Two days were sunny.

Six days were rainy.

Five days were cloudy.

Use the pictures to make a picture graph.

ONLINE EXPERIENCE You will click on an object and drag it to the spot it belongs.

HELPFUL HINT You can use S, R, or C to stand for each type of day.

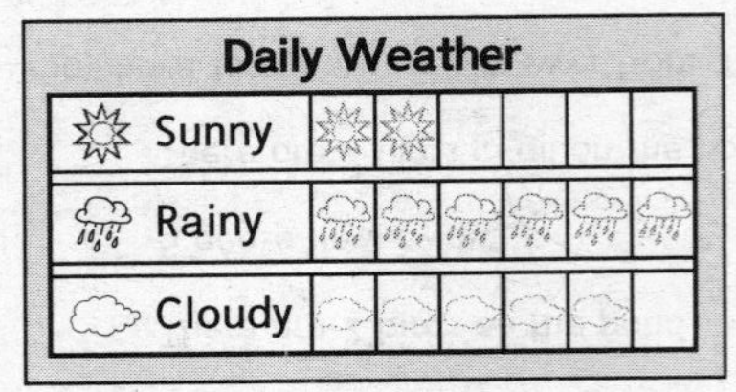

NAME ________ DATE ________

SCORE ________

Countdown: 20 Weeks

1. Which addition number sentence matches the picture? 1.OA.1

THINK SMART FOR SBAC
On the actual test, you might click on an answer to select it. In this book, you will circle the answer with a pencil.

A. 5 + 2 = 8
B. 4 + 3 = 7
C. 5 + 2 = 7 (circled)

2. Show a way to make the sum. Color some circles. Write the numbers that match the row of circles. 1.OA.1

Sample answer shown.

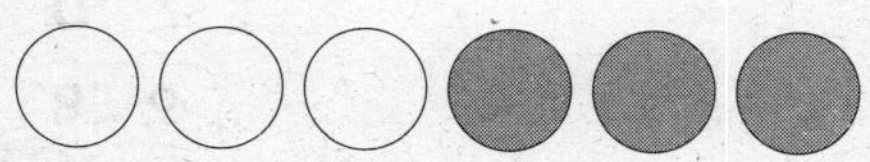

3 + 3 = 6

3. Find the whole. 1.OA.8

Part	Part
1	4
Whole	
5	

4. Add. 1.OA.6

$$\begin{array}{r} 2 \\ +\ 2 \\ \hline 4 \end{array}$$

5. Write an addition number sentence. Janessa picks 4 apples. Armani picks 5 more apples. How many apples do they have in all? 1.OA.1

4 + 5 = 9 apples

NAME ______________________ DATE ______________

SCORE ______________

Countdown: 19 Weeks

1. How many in all? 1.OA.1

A. 5

B. 6 (circled)

C. 7

2. Find the missing part of 10. 1.OA.8

Part	Part
2	8
Whole	
10	

THINK SMART FOR SBAC
On the actual test, you might be asked to use a keyboard to enter the number in the box. In this book, you will write the number with a pencil.

3. Add. 1.OA.6

2 + 5 = 7

4. True or false? 1.OA.1, 1.OA.7

Rolando finds 2 rocks at the beach. He finds 3 rocks in his yard. Rolando has 5 rocks.

True (circled) False

5. Write an addition sentence. 1.OA.1

6 + 2 = 8

NAME ______ DATE ______

SCORE ______

Countdown: 18 Weeks

1. A juggler is juggling 5 balls. He drops 2 balls. How many are left? 1.OA.1

3 balls

2. Subtract to fill in the missing part. 1.OA.1

Part	Part
1	6
Whole	
7	

THINK SMART FOR SBAC
On the actual test, you may be asked to click into the table and type your answer. In this book, you will write your answer in the table using a pencil instead.

3. There are 8 bugs crawling. 4 bugs stop. Write a subtraction number sentence to find the number of bugs still crawling. 1.OA.1

8 ⊖ 4 = 4 bugs

4. Subtract. 1.OA.3

$9 - 9 = 0$

5. There are 6 bunnies. 2 bunnies hop away. Cross out the bunnies to find how many bunnies are left. 1.OA.6

$$\begin{array}{r} 6 \\ -\ 2 \\ \hline 4 \end{array}$$

Copyright © McGraw-Hill Education. Permission is granted to reproduce for classroom use.

NAME ______ DATE ______

SCORE ______

Countdown: 17 Weeks

1. Write a subtraction sentence. 1.OA.6

6 – 3 = 3

2. Add. 1.OA.1

5 + 1 = 6

THINK SMART FOR SBAC
On the actual test, you may be asked to click on the blanks and type the numbers. In this book, you will write the numbers using a pencil instead.

3. There are 6 birds in the tree. There are 2 birds on the ground. Write a subtraction number sentence to find how many more birds there are in the tree than on the ground. 1.OA.1

6 ⊖ 2 = 4 birds

4. Subtract. 1.OA.6

5 – 2 = 3

5. There are 7 cars in the parking lot. 1 car drives away. How many cars are left? 1.OA.6

$$\begin{array}{r} 7 \\ -\ 1 \\ \hline 6 \end{array}$$

NAME DATE

SCORE

Countdown: 16 Weeks

THINK SMART FOR SBAC
On the actual test, you may be asked to select the correct answer using a mouse. In this book, you will circle the correct answer using a pencil.

1. Circle the number sentence that is false. I.OA.6

6 + 7 = 13

(8 + 9 = 18)

5 + 6 = 11

2. Jamal has 10 pennies in his left hand. He has 3 pennies in his right hand. How many pennies does Jamal have in all? I.OA.6

13 pennies

3. Use the number line to add. Write the sum. I.OA.6

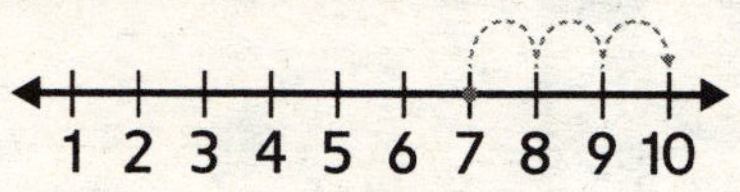

$$\begin{array}{r} 7 \\ +\ 3 \\ \hline 10 \end{array}$$

4. Mr. Heng has 9 buckets. He buys 9 more. How many buckets does Mr. Heng have in all? I.OA.6

18 buckets

5. Jamie has 9 cookies. She gives 3 away. How many cookies does Jamie have left? I.OA.7

6 cookies

NAME DATE

SCORE

Countdown: 15 Weeks

1. Tamika had 10 coins. She spent 4 of them. Tamika has 7 coins left. Circle true or false. 1.OA.7

   True (False)

> **THINK SMART FOR SBAC**
> On the actual test, you may be asked to select true or false by clicking on the words. In this book, you will use a pencil to draw a circle around the correct word.

2. There are 8 kids playing. 2 kids go home for dinner. How many kids are still playing? Write a subtraction number sentence. Then write the related addition fact. 1.OA.6

   8 − 2 = 6

   6 + 2 = 8

3. Circle the addition sentence that is true. 1.OA.7

   5 + 5 = 12

   (8 + 8 = 16)

   4 + 4 = 6

4. Pedro has 6 red blocks and 2 yellow blocks. How many blocks does Pedro have in all? Write two ways to add. 1.OA.3

   6 + 2 = 8 blocks

   2 + 6 = 8 blocks

5. Jenny saw 2 grasshoppers in the grass. She saw 3 grasshoppers on the sidewalk. She saw 1 grasshopper on the porch. How many grasshoppers did Jenny see in all? 1.OA.2

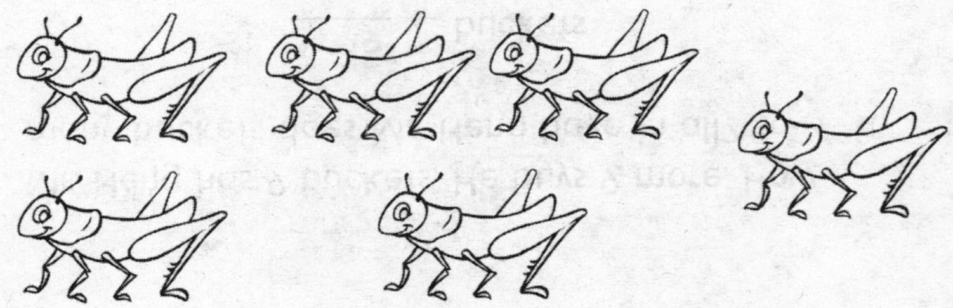

   6 grasshoppers

NAME ______ DATE ______

SCORE ______

Countdown: 14 Weeks

THINK SMART FOR SBAC
On the actual test, you may be asked to click into each blank to type the numbers. In this book, you will write the numbers using a pencil.

1. There are 11 robins in a tree. 3 of them fly away. How many robins are still in the tree? 1.OA.5

   11 – 3 = 8 robins

2. There are 13 balloons. 3 balloons pop. How many balloons are left? Use the number line to help you subtract. 1.OA.5

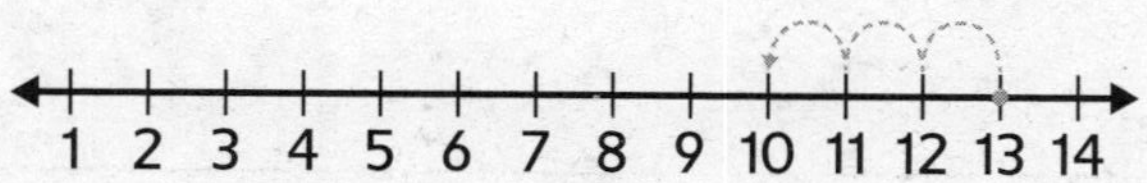

   10 balloons

3. Add or subtract. Draw lines to match the doubles facts. 1.OA.6

   3 + 3 = 6   8 – 4 = 4

   5 + 5 = 10   6 – 3 = 3

   4 + 4 = 8   10 – 5 = 5

4. Jasmine had 13 books. She gave 5 away. How many books does she have left? Write a number sentence to solve. 1.OA.6

   13 (–) 5 = 8 books

5. Winston has 5 rocks. He picks up 4 more. How many rocks does he have? 1.OA.6

$$\begin{array}{r} 5 \\ +\ 4 \\ \hline 9 \end{array}$$ rocks

NAME ______________________ DATE ______________

SCORE ______________

Countdown: 13 Weeks

1. There are 17 frogs on a log. 9 of the frogs hop in the pond. How many frogs are left? Take apart the number to make 10. Then subtract. 1.OA.6

17 – 9

7 | 2

17 – 7 = 10

10 – 2 = 8

17 – 9 = 8 frogs

2. Circle the number sentence that is not in the fact family. 1.OA.4

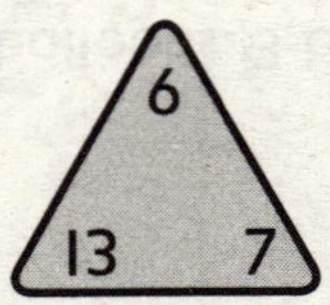

6 + 13 = 7

13 – 6 = 7

6 + 7 = 13

THINK SMART FOR SBAC
On the actual test, you may be asked to click on the facts to circle them. In this book, you will circle the answer using a pencil.

3. Andre sees 15 bugs. 6 of the bugs crawl away. How many bugs are left? Write an addition sentences and a subtraction sentence. 1.OA.6

6 + 9 = 15

15 – 6 = 9 bugs

4. Mr. Verne has 4 blacks hats and 2 brown hats. How many hats does Mr. Verne have in all? 1.OA.6

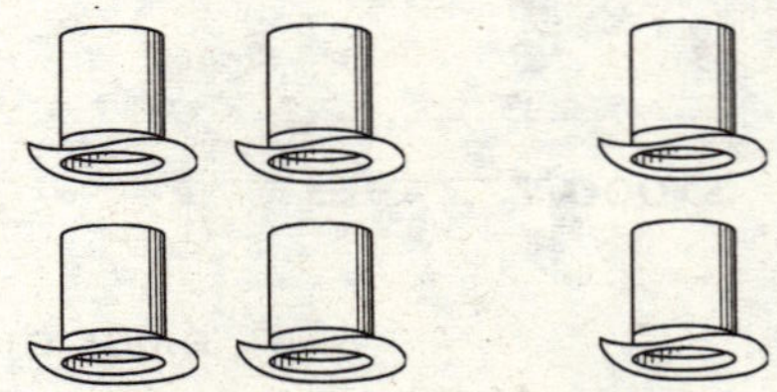

6 hats

5. Khrystian has 9 oranges and 9 apples. Khrystian has 18 pieces of fruit. Circle true or false. 1.OA.7

true   false

NAME ______________________ DATE ____________

SCORE ____________

Countdown: 12 Weeks

1. Mara has the peanuts shown. Circle a group of ten peanuts. Write how many more. How many peanuts does Mara have in all? 1.NBT.2b

> **THINK SMART FOR SBAC**
> On the actual test, you may be asked to circle the objects by clicking with a mouse. In this book, you will circle them using a pencil.

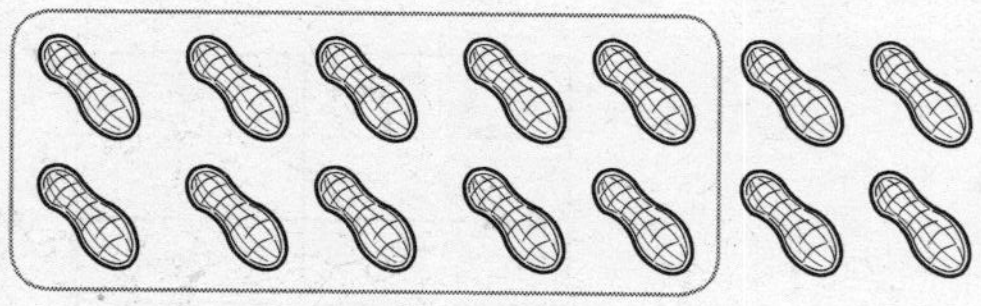

10 and __4__ more __14__ peanuts

2. Julia has 8 dimes. She spent 2 dimes. Count by tens. Write the numbers. How much does Julia have left? 1.NBT.1

__10__ ¢ __20__ ¢ __30__ ¢ __40__ ¢ __50__ ¢ __60__ ¢ left

3. There are six boxes. Each box has 10 toy cars in it. How many cars are there in all? 1.NBT.4c

__60__ cars

4. Pedro has 7 shelves of 10 books each. He also has 5 more books. How many books does he have in all? 1.NBT.2a

__75__ books

5. Circle the number sentence that is not a doubles fact. 1.OA.7

8 + 8 = 16

9 + 9 = 18

(7 + 7 = 13)

NAME ____________________ DATE ____________

SCORE ____________

Countdown: 11 Weeks

1. Lulu has 56 buttons. She has 5 sets of 10 green buttons. The rest are red. How many of the buttons are red? 1.NBT.2a

   6 red buttons

2. The Suarez family likes apples. They eat 10 apples every day. How many apples do they eat in 6 days. Fill in the table. 1.NBT.2c

   THINK SMART FOR SBAC
   On the actual test, you may be asked to click in the table cells and then type an answer. In this book, you will write your answers using a pencil.

Day	Apples
1	10
2	20
3	30
4	40
5	50
6	60

   60 apples

3. There are 4 groups of ten light bulbs. There are 3 extra light bulbs. How many light bulbs are there in all? 1.NBT.2a

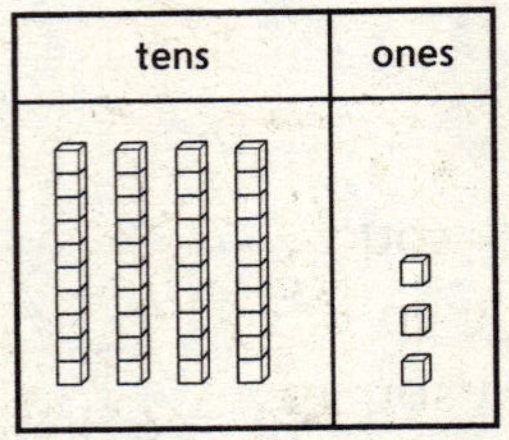

   43 light bulbs

4. When I am added to 6, the sum is 11. What number am I? 1.OA.4

   5

5. Henry found 3 small rocks. He also found 5 medium rocks and 4 big rocks. How many rocks did Henry find in all? 1.OA.2

   3 + 5 + 4 = 12 rocks

NAME ______________________ DATE ______________

SCORE ______________

Countdown: 10 Weeks

1. Are the number sentences part of the fact family? Circle yes or no. I.OA.6

> THINK SMART FOR SBAC
> On the actual test, you may be asked to select yes or no by clicking. In this book, you will instead circle the answer using a pencil.

(yes) no 7 + 5 = 12

(yes) no 5 + 7 = 12

yes (no) 7 − 5 = 12

(yes) no 12 − 5 = 7

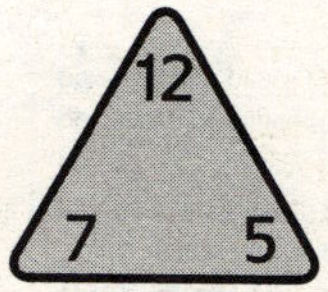

2. Alma's team has 36 points. The team scores 20 more points. How many points does the team have now? I.NBT.4

  36
+ 20
  56 points

3. There are 20 students in one class. There are 30 students in another class. How many students are there in all? I.NBT.4

50 students

4. There are 46 children at the beach. 3 more children come. How many children are at the beach in all? I.NBT.4

	tens	ones
	4	6
+		3
	4	9

49 children

5. Mrs. Gomez has two kinds of flowers. She has 13 flowers in all. Circle the kinds of flowers Mrs. Gomez has. I.NBT.4

(6 roses) 4 daisies (7 tulips)

Copyright © McGraw-Hill Education. Permission is granted to reproduce for classroom use.

Countdown

NAME ______________ DATE ______________

SCORE ______________

Countdown: 9 Weeks

1. Mr. Garmen has 17 chairs. He brings in 8 more. How many chairs does he have in all? Circle the ones to show regrouping. Write your answer. 1.NBT.4

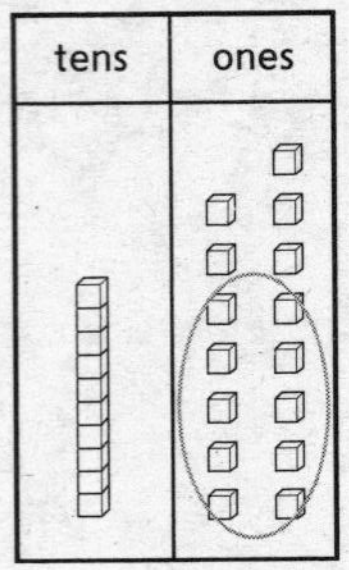

__25__ chairs

2. 40 + ______ = 80. Mark the related subtraction fact. 1.NBT.6

THINK SMART FOR SBAC
On the actual test, you may be asked to click into the boxes to select them. In this book, you will shade the boxes using a pencil instead.

- ☐ 40 + 10 = 80
- ■ 80 − 40 = 40
- ☐ 80 − 40 = 20
- ☐ 40 + 30 = 70

3. There are 50 players on a football team. 20 players leave the team. How many players are left? 1.NBT.6

__30__ players

4. Kylie had 47 stamps. She bought ten more stamps. Underline the number of stamps that Kylie has. 1.NBT.5

27 stamps 37 stamps

47 stamps <u>57 stamps</u>

5. Quentin buys a cookie that costs 45¢. He only used nickels to pay. How many nickels did Quentin use? 1.NBT.1

__9__ nickels

NAME ______________________ DATE ______________

SCORE ______________

Countdown: 8 Weeks

1. Write the totals. How many people like the color blue? I.MD.4

Favorite Colors		
Color	**Tally**	**Total**
Blue	卌 卌 ||	12
Green	卌 |||	8
Yellow	||||	4

12 people like blue.

2. George asks his friends to name their favorite sport. 4 people like football. 2 people like baseball. 3 people like soccer. Make a graph. I.MD.4

THINK SMART FOR SBAC On the actual test, you may be asked to drag the pictures into the graph in order to tell how many. In this book, you will draw the pictures using a pencil.

3. Which kind of plant has more votes than flowers? I.MD.4

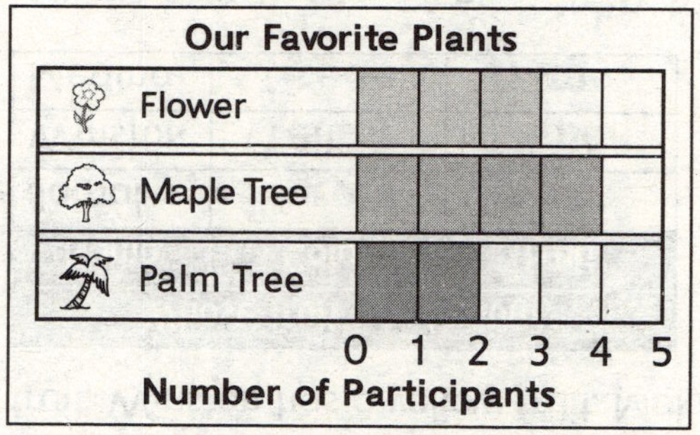

Answer: Maple Tree

4. Rayshawn has 7 small cars and 6 big cars. Rayshawn has 14 cars. Circle true or false. I.OA.7

true (false)

5. Write each number. Circle *is greater than, is less than,* or *is equal to.* I.NBT.3

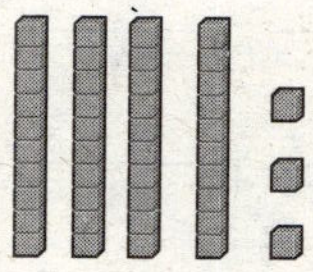

(is greater than)
is less than
is equal to

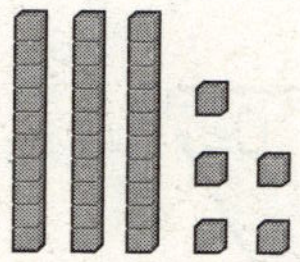

43 ______ 35 ______

Countdown

NAME _______________ DATE _______________

SCORE _______________

Countdown: 7 Weeks

1. Write the tallies. What flavor do people like the most? I.MD.4

Favorite Flavors		
Flavor	Tally	Total
Vanilla	卌 I	6
Chocolate	卌 II	7
Strawberry	卌 IIII	9

THINK SMART FOR SBAC
On the actual test, you may be asked to drag the tally marks into the table using a mouse. In this book, you will draw the tally marks using a pencil.

People like strawberry the most.

2. Mercedes is counting toys. She counts 12 toys in all. How many dolls does she count? I.MD.4

3 dolls

3. Write the numbers in to make a fact family. I.NBT.6

20 + 30 = 50

30 + 20 = 50

50 – 20 = 30

50 – 30 = 20

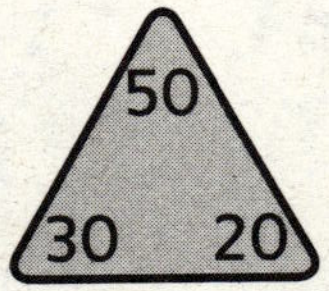

4. There are 12 fish near a rock. 2 of them swim away. How many fish are left? I.OA.5

12 – 2 = 10 fish

5. Jamal, Winston, and Medina each have a piece of fruit. The fruits are a lime, a lemon, and a plum. Jamal has a purple fruit. Medina has a green fruit. Winston has a yellow fruit. Make a table. I.MD.4

Whose fruit is a lemon?		
Name	Color	Fruit
Jamal	Purple	Plum
Winston	Yellow	Lemon
Medina	Green	Lime

Whose fruit is a lemon? Winston

NAME ______________________ DATE __________

SCORE __________

Countdown: 6 Weeks

1. A skateboard is shorter than a bike. A bike is shorter than a car. Is a car longer than or shorter than a skateboard? Circle the answer. I.MD.I

shorter than   (longer than)

2. Naomi has three brushes. Match each brush to the number. I is for short. 2 is for shorter. 3 is for shortest. I.MD.I

THINK SMART FOR SBAC
On the actual test, you may be asked to draw lines using a mouse. In this book, you will draw the lines using a pencil.

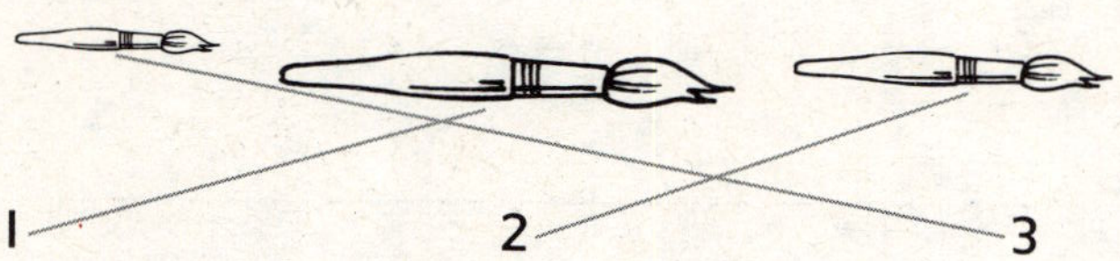

3. A pea plant was 4 cubes high on Monday. It grows 2 cubes higher every day. How many cubes high is it on Wednesday? Complete the table. I.MD.2

Days	Cubes
Monday	4
Tuesday	6
Wednesday	8

8 cubes high

4. Mr. Uher gets home at 5 o'clock. Mrs. Uher gets home one hour earlier. What time does Mrs. Uher get home? I.MD.3

4 o'clock

5. There are 40 oranges in one basket. Geraldine puts 30 more oranges in. How many oranges are in the basket? I.NBT.4

70 oranges

NAME ______________________ DATE ______________________

Countdown: 5 Weeks

SCORE ______________________

THINK SMART FOR SBAC
On the actual test, you may be asked to drag the hands around the clock to mark the correct time. In this book, you will draw the hands using a pencil.

1. Kiane eats dinner at 6 o'clock. She starts reading one hour later. Mark the time on the clock that Kiane starts reading. I.MD.3

2. Liam finished running at the time on the clock. He ran for I hour. What time did Liam start running? I.MD.3

<u>I</u> o'clock

3. Yolanda's mom picks her up from school at 3:00. They drive 30 minutes home. What time do they get home? I.MD.3

<u>3:30</u>

4. Add or subtract. Draw lines to match the doubles facts. I.OA.6

7 + 7 = <u>14</u>	14 − 7 = <u>7</u>
9 + 9 = <u>18</u>	12 − 6 = <u>6</u>
6 + 6 = <u>12</u>	18 − 9 = <u>9</u>

5. A store has 50 toy cars. They sell 40 of the cars. How many toys cars are left? Use the number line to help. I.NBT.6

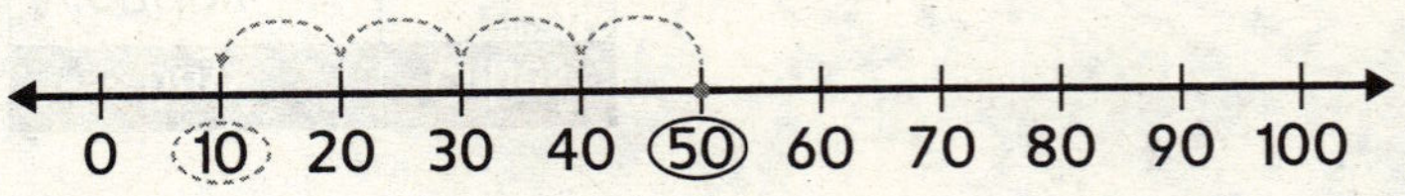

<u>10</u> toy cars

NAME ________ DATE ________

SCORE ________

Countdown: 4 Weeks

1. Circle the object that has a square. I.G.I

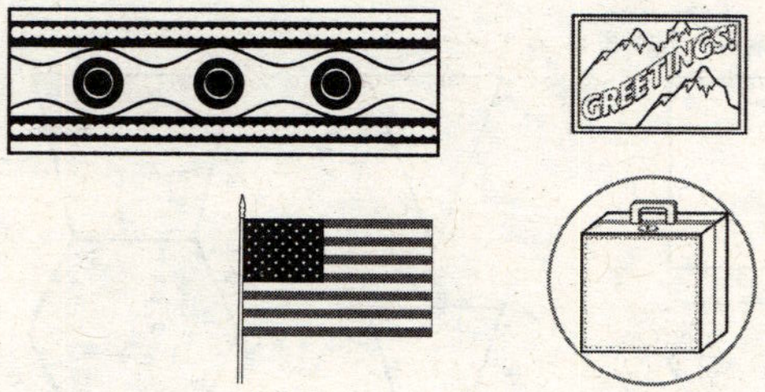

2. Draw a line in the picture to make a triangle and a trapezoid. I.G.I

THINK SMART FOR SBAC
On the actual test, you may be asked to draw the line using a mouse. In this book, you will draw the line using a pencil.

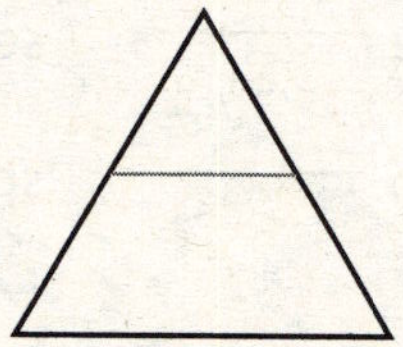

Sample answer. Any line parallel to one side of the triangle is correct.

3. Circle the words that describe the shape. I.G.I

(trapezoid) triangle (closed)

3 vertices (4 sides) 0 vertices

4. Francisco wants to cover the hexagon with 4 shapes. How many of each shape does he need to make the hexagon? I.G.2

1

3

5. Josiah has practice at half past 3. Practice lasts for one hour. At what time is practice over? Write the time on the clock. I.MD.3

NAME ______________ DATE ______________

SCORE ______________

Countdown: 3 Weeks

1. Count how many of each shape. I.G.1

2 trapezoids 2 circles

3 rectangles 3 triangles

2. Circle the shapes that can be used to make the design. I.G.2

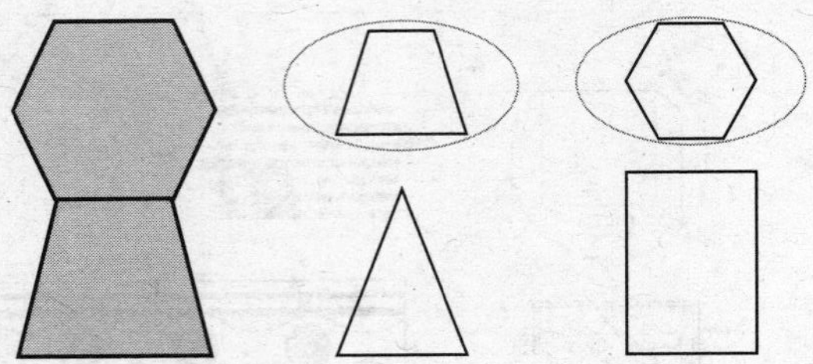

THINK SMART FOR SBAC
On the actual test, you may be asked to circle the shapes by clicking on them. In this book, you will draw the circles using a pencil.

3. Mrs. Chin bakes a cake. She cuts it into 4 equal parts. Circle the pictures that could be Mrs. Chin's cake. I.G.3

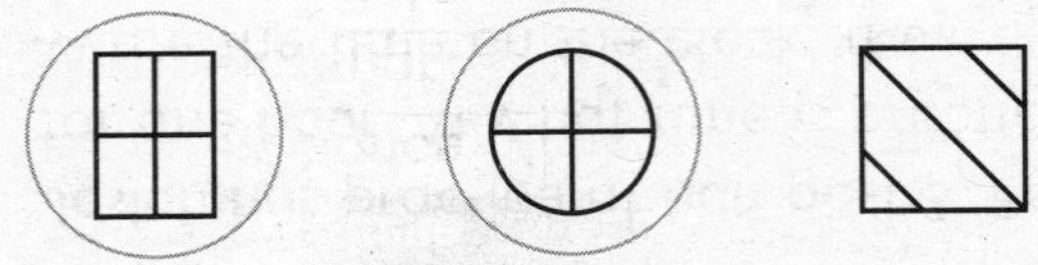

4. It is half past the hour. The hour hand is between 3 and 4. What time is it? Circle the correct time. I.MD.3

2:30 3:30 4:30

5. Subtract to fill in the missing part. I.OA.1

Part	Part
10	5
Whole	
15	

Copyright © McGraw-Hill Education. Permission is granted to reproduce for classroom use.

NAME ________________ DATE ________

SCORE ________

Countdown: 2 Weeks

1. What is the name of the shape that makes up the faces of the object? I.G.1

square

2. Circle the names of the shapes that are not used to make the solid. I.G.2

THINK SMART FOR SBAC
On the actual test, you may be asked to click on the words to select them. In this book, you will draw the circles using a pencil.

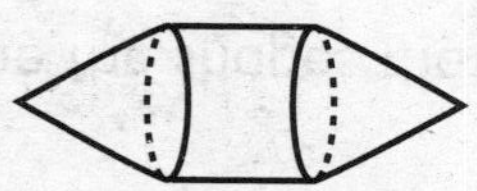

cone — circle

cylinder — rectangular prism

triangle

3. A fish tank has six rectangular faces. What shape is the fish tank? I.G.1

rectangular prism

4. Mr. Fan has ice cream that is served in a cone. The cone has a face that is a square. Circle true or false. I.G.1

true — false

5. Chance, Lamar, and Gretchen each have a cup of yogurt. The flavors are vanilla, chocolate, and strawberry. Chance has brown yogurt. Gretchen has white yogurt. Lamar has pink yogurt. Make a table. I.MD.4

Whose flavor is chocolate?		
Name	Color	Flavor
Chance	Brown	Chocolate
Lamar	Pink	Strawberry
Gretchen	White	Vanilla

Whose flavor is chocolate? Chance

NAME DATE

SCORE

Countdown: I Week

1. Garrett made a pattern with blocks. What two shapes does he need to complete his pattern? I.G.1

(cone) rectangular prism (cylinder)

2. Underline the shapes that have six faces. I.G.1

cone cylinder

cube rectangular prism

THINK SMART FOR SBAC
On the actual, test you may be asked to underline the words using a mouse. In this book, you will underline the words using a pencil.

3. Juan is counting fruit. I.MD.4

How many more oranges than apples are there? 3

4. Isabella wants to divide a trapezoid into a rectangle and two triangles. Draw the lines to help Isabella. I.G.2

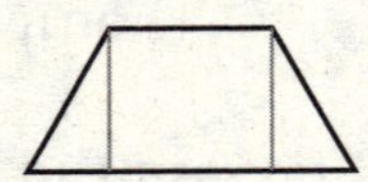

5. A can of soup has 2 faces shaped like circles and no vertices. What is the name of the shape? I.G.1

cylinder

NAME ______________ DATE ______________

SCORE ______________

Chapter 1 Test

1. Find the matching addition number sentence. 1.OA.1

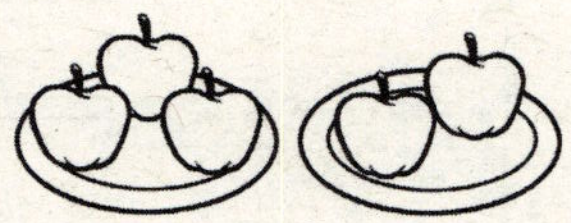

A. 3 + 2 = 5 (circled)

B. 2 + 3 = 6

C. 1 + 4 = 5

2. Find the missing part. 1.OA.1

Part	Part
7	2
Whole	
9	

3. Write an addition number sentence. 1.OA.1

3 + 5 = 8 turtles

4. Add zero. 1.OA.3

9 + 0 = 9

5. True or false?

Jeremiah has 5 toy cars. He gets 2 more for his birthday. Jeremiah has 6 cars in all. 1.OA.1, 1.OA.7

True False (circled)

6. True or false?

Fernando has 2 books on one shelf and 3 books on another shelf. Fernando has 6 books in all. I.OA.6, I.OA.7

True ~~~ **(False)** (circled)

7. Add. I.OA.6

$4 + 3 = \underline{7}$

8. Add. I.OA.6

$5 + 1 = \underline{6}$

9. True or false? I.OA.6, I.OA.7

$2 + 4 = 6$

(True) (circled) ~~~ False

10. Add. I.OA.6

$$\begin{array}{r} 5 \\ +\ 5 \\ \hline 10 \end{array}$$

11. Add. I.OA.6

$$\begin{array}{r} 1 \\ +\ 8 \\ \hline 9 \end{array}$$

12. Find the missing part. 1.OA.6

Part	Part
6	4
Whole	
10	

13. Geraldine has 5 fish. Circle the ways that will make 5 fish. 1.OA.6

(2 + 3) 4 + 2

(1 + 4) (0 + 5)

14. Mr. Bake has 4 brown chairs and 4 black chairs. How many chairs does Mr. Bake have in all? 1.OA.6

8 chairs

15. Circle two numbers that will make 10. 1.OA.6

(2) 6 3

0 (8) 5

Chapter Tests

NAME ______________ DATE ______________

SCORE ______________

Chapter 2 Test

1. Find the matching subtraction number sentence. I.OA.1, I.OA.5

A. 8 − 5 = 3

B. 9 − 6 = 3

C. 9 − 3 = 6

2. Find the missing part. I.OA.4, I.OA.8

Part	Part
2	4
Whole	
6	

Subtract. I.OA.6

3. 5 − 2 = 3

4. 10 − 2 = 8

5. True or false? I.OA.1, I.OA.7

Zoe has 6 books. She takes 2 back to the library. Now Zoe has 5 books.

True False

6. Write a subtraction number sentence. I.OA.I

9 – 2 = 7

7. Subtract 0. I.OA.6, I.OA.8

5 – 0 = 5

8. True or false? I.OA.7

4 – 3 = 1

(True) False

Subtract. I.OA.6

9. 10 – 5 = 5

10. 7 – 6 = 1

11. 2 – 0 = 2

12. Write the related subtraction fact. 1.OA.6

8 − 3 = 5

8 − 5 = 3

13. Find the missing part. 1.OA.4, 1.OA.8

Part	Part
6	4
Whole	
10	

A. 4

B. 3

C. 5

14. Write a subtraction number sentence. 1.OA.1

Xavier has 5 cheese cubes. He eats 3 of them. How many cheese cubes are left?

5 − 3 = 2

15. Find the related subtraction number sentence. 1.OA.6

10 − 3 = 7

A. 7 − 4 = 3

B. 10 − 6 = 4

C. 10 − 7 = 3

NAME ____________ DATE ____________

Chapter 3 Test

SCORE ____________

1. Chase starts at the number 9. He counts "9, 10, 11, 12." Circle the number sentence that matches Chase's counting. 1.OA.5

   (9 + 3 = 12)

   10 + 2 = 12

   9 + 4 = 12

2. Xavier has 7 pennies in his left pocket and 6 pennies in his right pocket. How many pennies does Xavier have in all? 1.OA.6

   7¢ + 6¢ = 13 ¢

3. Use the number line to add. Write the sum. 1.OA.6

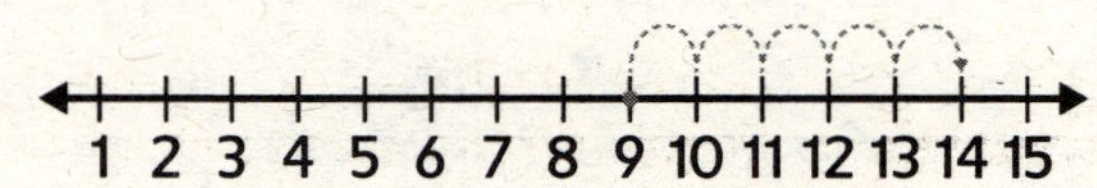

$$\begin{array}{r} 5 \\ +\ 9 \\ \hline 14 \end{array}$$

4. Circle the correct doubles facts. 1.OA.6

   5 + 5 = 12   6 + 6 = 10

   (4 + 4 = 8)   (7 + 7 = 14)

5. Shalyn took 8 pictures on Friday. She took 5 more pictures on Saturday. How many pictures did she take in all? 1.OA.6

   13 pictures

6. Javier's teacher asked him to add 8 + 4. He makes 10 to add. Which number sentence does Javier use? 1.OA.6

10 + 1 = 11

(circled) 10 + 2 = 12

10 + 3 = 13

7. Darian has 5 baseball cards and 9 football cars. How many cards does Darian have in all? Write two ways to add. 1.OA.3

__5__ + __9__ = __14__ cards

__9__ + __5__ = __14__ cards

8. Christina bought 3 apples, 5 oranges, and 9 bananas. How many pieces of fruit did she buy in all? 1.OA.2

__3__ + __5__ + __9__ = __17__ pieces

9. True or false? 1.OA.2, 1.OA.7

4 + 3 + 2 = 10

True (circled) False

10. Which sum equals 20? 1.OA.2

A. 5 + 6 + 7

B. 2 + 7 + 8

(circled) C. 6 + 6 + 8

11. True or false?

Bryn has 6 white flowers and 7 red flowers. Her sister has 7 blue flowers and 6 pink flowers. Bryn and her sister have the same number of flowers. 1.OA.3, 1.OA.7

(True) False

12. Make 10 to add 8 + 6. 1.OA.6

10 + 4 = 14

13. Circle any sums that equal 15. 1.OA.6

(10 + 5) (9 + 6)

(8 + 7) 6 + 4

14. Veronica saw 9 birds in her front yard and 9 birds in her back yard. How many birds did she see in all? 1.OA.6

18 birds

15. Kerin's team scored 8 points. The team scored 6 more points. How many points did they have in all? 1.OA.6

14 points

Copyright © McGraw-Hill Education. Permission is granted to reproduce for classroom use.

NAME ________________ DATE ________

SCORE ________

Chapter 4 Test

1. Derrick starts counting backwards from 19. He says, "19, 18, 17, 16." Circle the number sentence he is trying to solve. 1.OA.5

   (19 − 3 = 16) (circled)

   19 − 4 = 16

   19 − 4 = 15

2. There are 12 muffins. The Perez family eats 5 of them. How many muffins are left? Use the number line to help you subtract. 1.OA.5

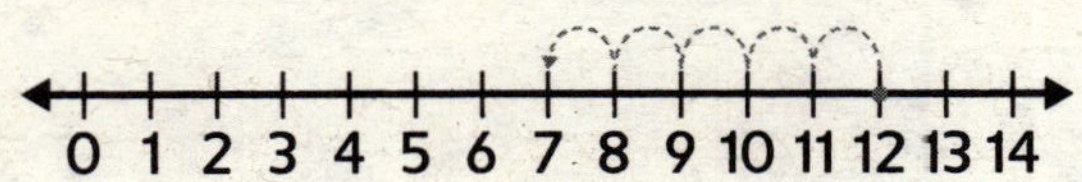

   __7__ muffins

3. Add or subtract. Draw lines to match the doubles facts. 1.OA.6

   6 + 6 = __12__ 14 − 7 = __7__

   7 + 7 = __14__ 12 − 6 = __6__

   8 + 8 = __16__ 16 − 8 = __8__

4. There are 18 bats in the zoo. 9 are sleeping. How many bats are awake? 1.OA.6

   __9__ bats

5. Huan solved 13 math problems. He got 4 incorrect. How many problems did Huan get correct? Write a number sentence to solve. 1.OA.1

   __13__ (−) __4__ = __9__ problems

6. Write a subtraction sentence for the following number line. 1.OA.5

13 – 9 = 4

7. There are 15 elephants drinking at the spring. 7 elephants leave. How many elephants are left? Take apart the number to make 10. Then subtract. 1.OA.6

15 – 7
5 2

15 – 5 = 10

10 – 2 = 8

15 – 7 = 8 elephants

8. Iesha checked out 14 books from the library. She read 8 of the books. How many books did she not read? Write an addition sentence and a subtraction sentence. 1.OA.6

6 + 8 = 14

14 – 8 = 6 books

9. Which subtraction fact is not related to 5 + 6 = 11? 1.OA.6

A. 11 – 5 = 6

B. 11 – 6 = 5

C. 6 – 11 = 5

10. Circle the number sentence that is not in the fact family. 1.OA.4

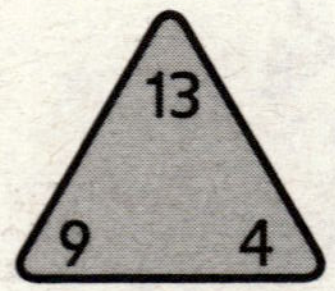

4 + 9 = 13

(circled) 9 + 13 = 4

13 − 9 = 4

11. Maria is thinking of a number. When she adds 6 to the number, she gets 10. What is her number? 1.OA.4

6 + __4__ = 10

12. Marquis has 6 shoes and some socks. He has 14 shoes and socks in all. How many socks does Marquis have? 1.OA.4

__8__ socks

13. True or false?

The two number sentences are in the same fact family. 1.OA.4

9 + 9 = 18

18 − 9 = 9

(circled) True   False

14. Which subtraction problem has the same answer as 12 − 5? 1.OA.6

10 − 2

(circled) 10 − 3

10 − 4

15. Lena listened to 15 songs. She liked 10 of the songs. How many songs did Lena not like? 1.OA.4

10 + __5__ = 15

__5__ songs

NAME DATE

SCORE

Chapter 5 Test

1. Mrs. Browning has the scissors shown. Circle a group of ten scissors. Write how many more. How many scissors does Mrs. Browning have in all? I.NBT.2b

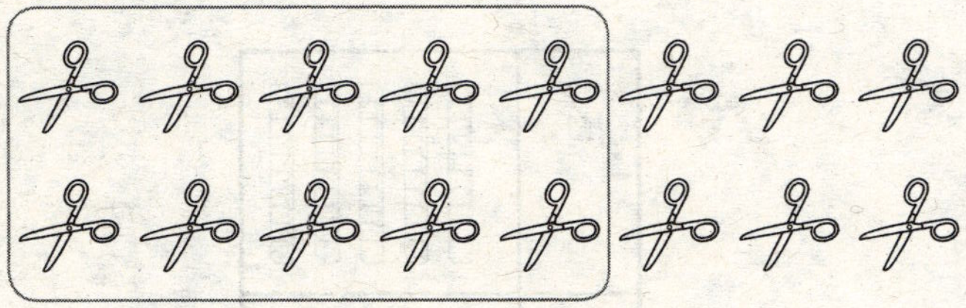

10 and __6__ more __16__ scissors

2. Gwen has 10 white flowers and 8 red flowers. How many flowers does she have in all? I.NBT.2c

__18__ flowers

3. Eunice has 40 baskets. She gets 2 more. How many baskets does Eunice have? I.NBT.2a

A. 42 baskets (circled)

B. 24 baskets

C. 44 baskets

4. Tisha has 76 ribbons. How many groups of ten does she have? How many ones? I.NBT.2a

__7__ tens and __6__ ones

5. True or false? George has 56 marbles. Samuel has 65 marbles. Samuel has ten more marbles than George. I.NBT.5

True False (circled)

6. Jackson had a lemonade stand. Count by tens to find out how much money Jackson made. I.NBT.1

10 ¢ 20 ¢ 30 ¢ 40 ¢ 50 ¢ 60 ¢ 70 ¢ in all

7. There are 5 groups of ten cars. There are 2 extra cars. How many cars are there in all? I.NBT.2a

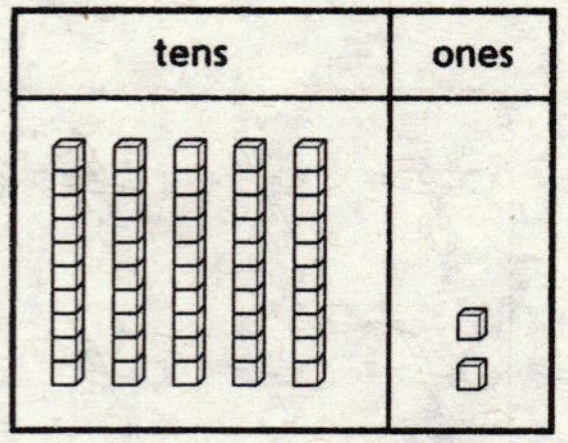

52 cars

8. Jorge buys a bracelet that costs 75¢. He only used nickels to pay. How many nickels did Jorge use? I.NBT.1

15 nickels

9. Circle the right words. I.NBT.3

46 is less than (circled) / is greater than / is equal to 64

10. On vacation Ms. Hernandez took pictures. She took 1 hundred, 1 ten, and 6 ones. How many pictures did Ms. Hernandez take? I.NBT.1

116 pictures

11. Mr. Akita buys 10 stamps every week. How many stamps does he buy in 5 weeks? Fill in the table. 1.NBT.2c

Week	Stamps
1	10
2	20
3	30
4	40
5	50

50 stamps

12. Tory is thinking of a number. It is between 109 and 111. What is her number? 1.NBT.1

A. 108

B. 110

C. 112

13. Seamus has 35 nickels. Miranda has 41 nickels. Write <, >, or =. Tell who has more. 1.NBT.3

35 (<) 41

Who has more? Miranda

14. Naomi picked eighty-seven apples. Circle the number of apples Naomi picked? 1.OA.7

87

78

18

15. True or false?

7 tens and 8 ones is the same as 8 tens and 7 ones. 1.NBT.5

True False

NAME ________ DATE ________

Chapter 6 Test

SCORE ________

1. There are 50 players on one team. There are 30 players on another team. How many players are there in all? I.NBT.4

   80 players

2. Circle the correct answer. I.NBT.4

$$\begin{array}{r} 40 \\ +\ 40 \\ \hline \end{array}$$

   A. 8

   B. 48

   C. 80

3. Gina sells 24 glasses of lemonade. She sells 20 more glasses. How many glasses does Gina sell in all? I.NBT.4

$$\begin{array}{r} 24 \\ +\ 20 \\ \hline 44 \end{array}$$ glasses

4. Add. I.NBT.4

$$\begin{array}{r} 45 \\ +\ 3 \\ \hline 48 \end{array}$$

5. Add. I.NBT.4

$$\begin{array}{r} 5 \\ +\ 62 \\ \hline 67 \end{array}$$

6. Sammy saw two kinds of bugs. He saw 16 bugs in all. Circle the kinds of bugs Sammy saw. I.NBT.4

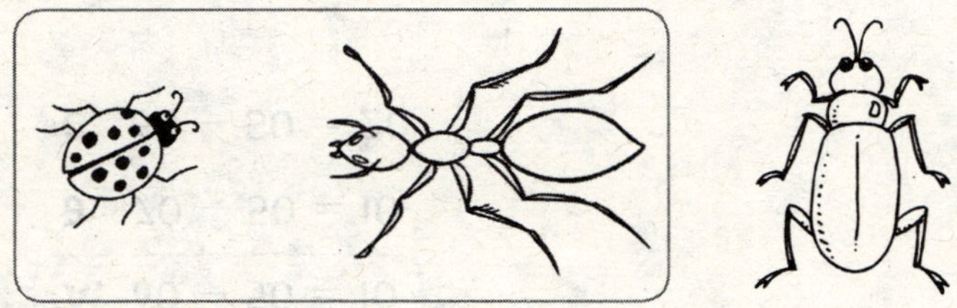

5 ladybugs II ants 6 beetles

7. Mr. Shimp has 26 pens. He buys 7 more. How many pens does he have in all? Circle the ones to show regrouping. Write your answer. I.NBT.4

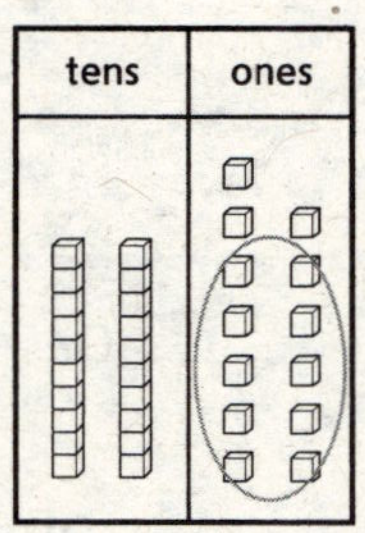

33 pens

8. True or false? I.NBT.4

$$\begin{array}{r} 45 \\ +\ 7 \\ \hline 412 \end{array}$$

True False

9. Kevin counted some cars. He counted 60 cars in all. 20 cars were red. The others were blue. How many blue cars did Kevin count? I.NBT.6

40 blue cars

10. Use the number line to subtract. Write the difference. I.NBT.6

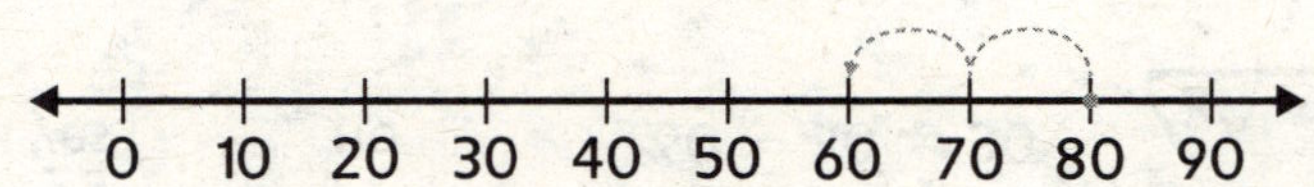

80 − 20 = 60

Chapter Tests

11. 30 + ______ = 70. Mark the related subtraction fact. I.NBT.6

- [x] 70 − 30 = 40
- [ ] 70 − 30 = 10
- [ ] 40 + 30 = 70

12. Circle the number sentence that matches the number line. I.NBT.6

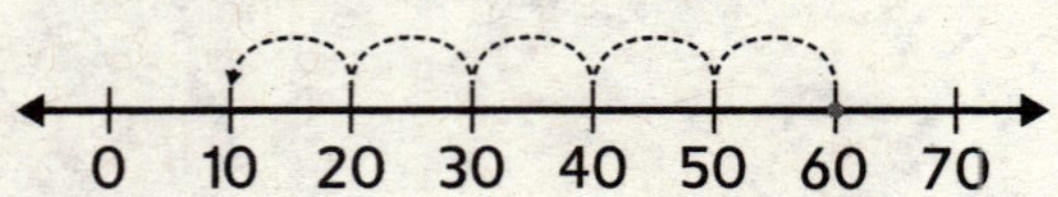

A. 60 − 50 = 10 (circled)

B. 70 − 50 = 10

C. 60 − 50 = 20

13. Are the number sentences part of the fact family? Circle yes or no. I.NBT.6

yes (circled)	no	30 + 30 = 60
yes (circled)	no	60 − 30 = 30
yes	no (circled)	6 − 3 = 3

14. Aaron hit 23 homeruns in his first season. In his second season he hit 20 homeruns. How many homeruns did Aaron hit in all? I.NBT.4

__43__ homeruns

15. True or false?

76 is 70 more than 6. I.NBT.4

True (circled) False

NAME ____________ DATE ____________

Chapter 7 Test

SCORE ____________

1. Write the totals. I.MD.4

Favorite Season		
Season	Tally	Total
Winter	卌 ||||	9
Spring	卌 卌	10
Summer	||	2
Autumn	卌 卌 |	11

How many more people like Winter than Summer?

7 people

2. Henry, Chloe, and Jose each have a pet. The pets are a dog, a cat, and a bird. Henry has the pet that barks. Jose has the pet that tweets. Make a table. I.MD.4

Whose pet is a cat?		
Name	Sound	Pet
Henry	Bark	Dog
Chloe	Meow	Cat
Jose	Tweet	Bird

Whose pet is the cat? Chloe

3. Write the tallies. What sport do people like the most? I.MD.4

Favorite Sports		
Sport	Tally	Total
Football	卌 ||	7
Soccer	卌 卌	10
Tennis	卌	5

People like soccer the most.

4. Arianna asks her friends to name their favorite donut. 7 people like plain. 3 people like iced. 2 people like filled. Make a graph. I.MD.4

5. How many animals are there in all? I.MD.4

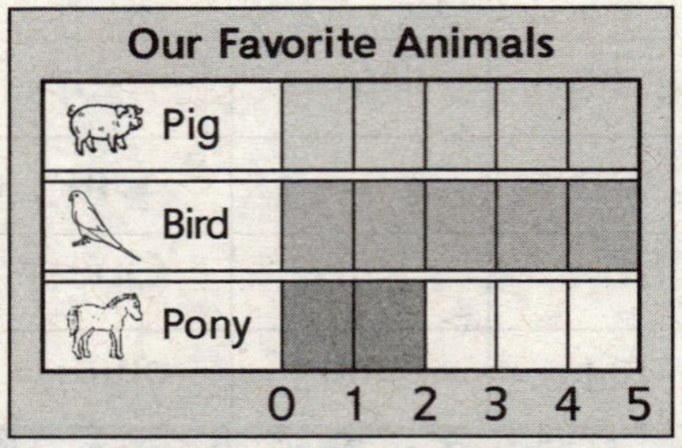

12 animals

6. Mr. Johnson, Ms. Baird, and Mrs. Blackthorne each teach a math class. The sizes of the classes are 9, 15, and 22. Mrs. Blackthorne has the class with fewer than 10 students. Mr. Johnson has the class with more than 20 students. Make a table. I.MD.4

Whose class has 15 students?	
Name	How many students?
Mr. Johnson	22
Ms. Baird	15
Mrs. Blackthorne	9

Whose class has 15 students? Ms. Baird

7. Mrs. Tyler has 5 plates, 8 forks, and 7 spoons. Make a bar graph. I.MD.4

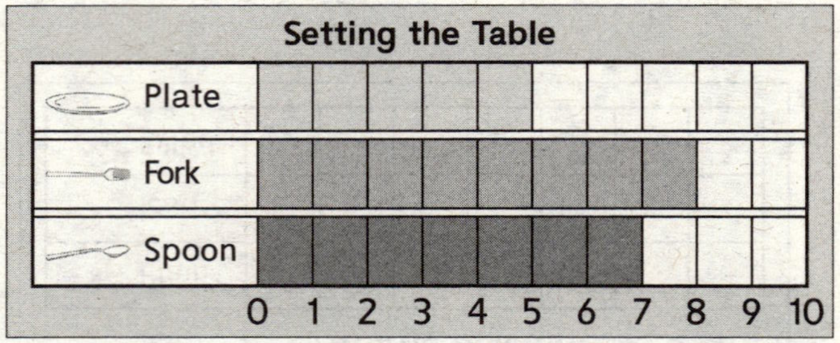

8. Look at the bar graph for the number of people who like each holiday. Make a tally chart. I.MD.4

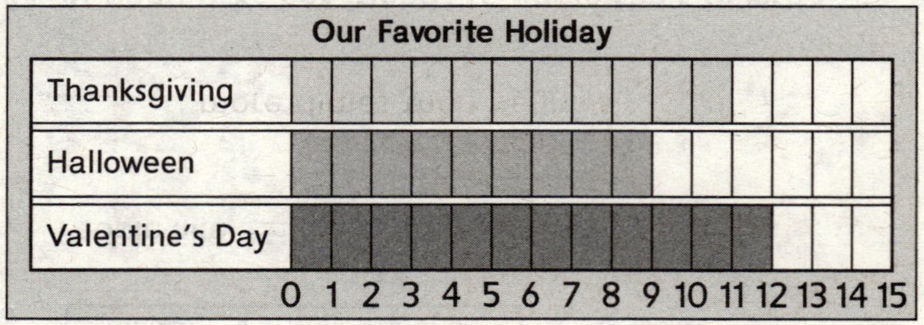

Favorite Holiday		
Holiday	Tally	Total
Thanksgiving	𝍸 𝍸 I	11
Halloween	𝍸 IIII	9
Valentine's Day	𝍸 𝍸 II	12

9. Two teams scores 27 points in all. The Blue Jays score 14 points. Make a tally chart to show how many points each team scores. I.MD.4

Points Scored		
Points	Tally	Total
Blue Jays	~~||||~~ ~~||||~~ ||||	14
Cardinals	~~||||~~ ~~||||~~ |||	13

10. Francine is counting animals in a pond. She counts 15 animals in all. How many ducks does she count? I.MD.4

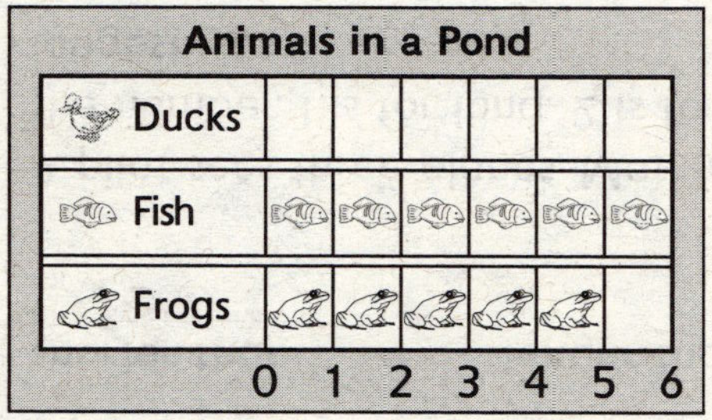

4 ducks

11. Camden collected coins during a lemonade sale. How many more dimes did he collect than pennies? I.MD.4

3 more dimes than pennies

12. Which kind of hat has more votes than the baseball cap? I.MD.4

Answer: Top Hat

NAME ______________ DATE ______________

Chapter 8 Test

SCORE ______________

1. The table is longer than the chair. The carpet is longer than the table. Is the chair longer than or shorter than the carpet? Circle the answer. I.MD.I

shorter than (circled)   longer than

2. A pilot sees three planes. Match each plane to the number. I is for long. 2 is for longer. 3 is for longest. I.MD.I

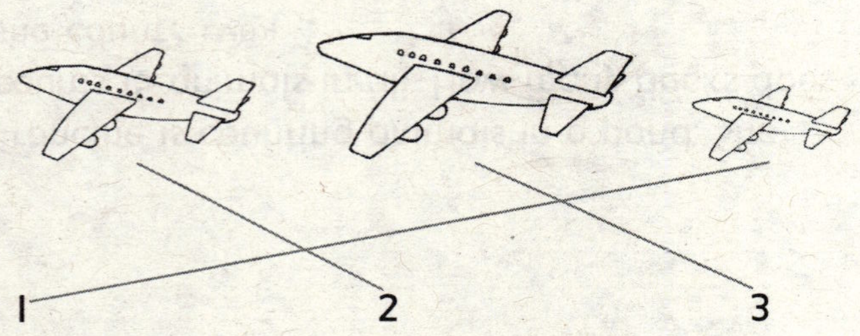

3. How many acorns long is the caterpillar? I.MD.2

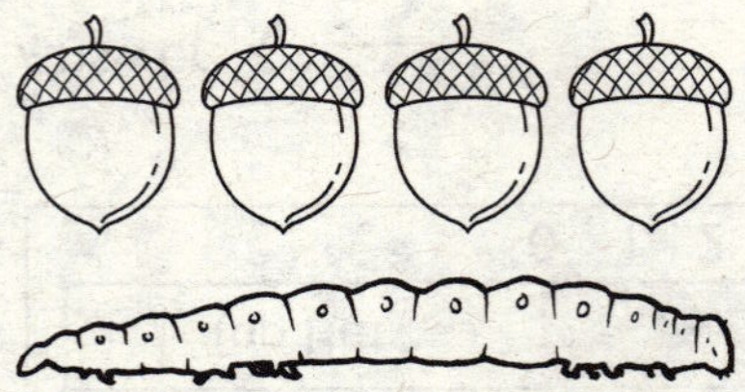

4 acorns

4. A corn plant was 7 cubes high on Thursday. It grows 3 cubes higher every day. How many cubes high is it on Sunday? I.MD.2

Days	Cubes
Thursday	7
Friday	10
Saturday	13
Sunday	16

16 cubes high

5. Match the times on the clocks. I.MD.3

6. It is half past the hour. The hour hand is between 7 and 6. What time is it? Circle the correct time. I.MD.3

5:30   (6:30)   7:30

7. Harry started his homework at 3 o'clock. He got home from school 1 hour earlier. What time did Harry get home? I.MD.3

___2___ o'clock

8. Pierre's mom drives to the mall at 4:30. She drives 30 minutes. What time does she arrive at the mall? I.MD.3

___5:00___

9. The pencil is shorter than the piece of paper. The piece of paper is shorter than the book. Which object is the longest? I.MD.1

A. The pencil

B. The piece of paper

C. The book (circled)

10. Amal had 17 cookies. He eats two cookies a day. How many cookies does he have left on Day 4? I.MD.2

Days	Cookies
Day 1	17
Day 2	15
Day 3	13
Day 4	11

11 cookies

11. Geraldine wakes up at 7 o'clock. She eats breakfast one hour later. Mark the time on the clock that Geraldine starts eating breakfast. I.MD.3

12. The Ramirez family leaves for the fair at the time shown on the clock. They arrive 2 hours later. Mark the time on the other clock that shows what time they arrived. I.MD.3

13. Martha has practice at half past noon. Practice lasts for one hour. What time is practice over? I.MD.3

14. Circle the clock that shows the latest time. I.MD.3

NAME ______________________ DATE ______________

Chapter 9 Test

SCORE ______________

1. Count how many of each shape. I.G.1

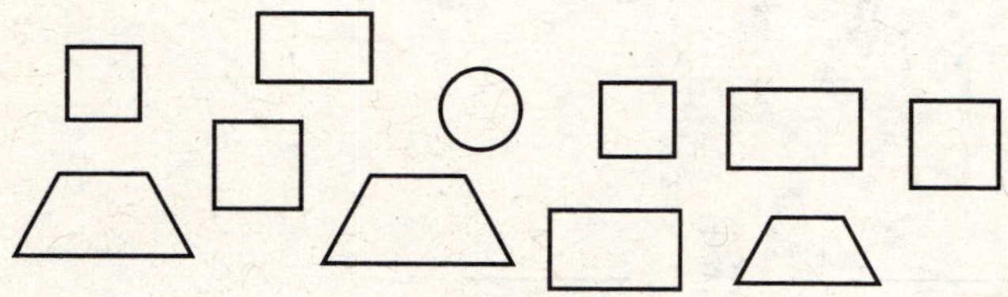

__3__ trapezoids __1__ circles

__3 or 7__ rectangles __4__ squares

2. Circle the shapes that can be used to make the design. I.G.2

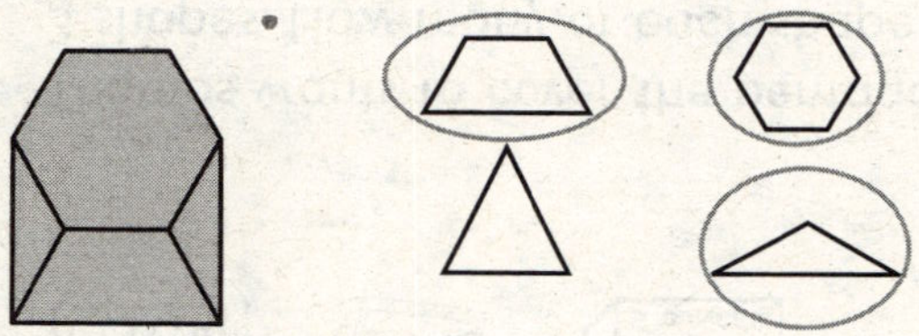

3. Circle all the words that describe the shape. I.G.1

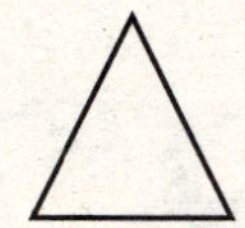

trapezoid (triangle) (closed)

(3 vertices) 4 sides 0 vertices

4. Draw two lines in the picture to make a rectangle and two triangles. I.G.1

Sample answer is shown.

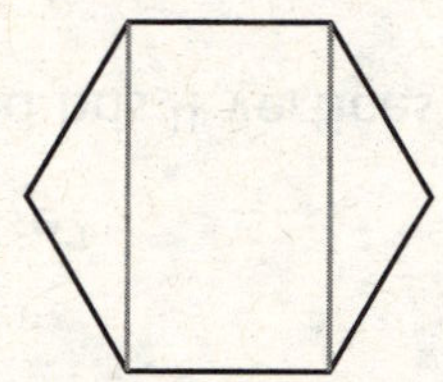

5. Circle the object that has a square in it. I.G.1

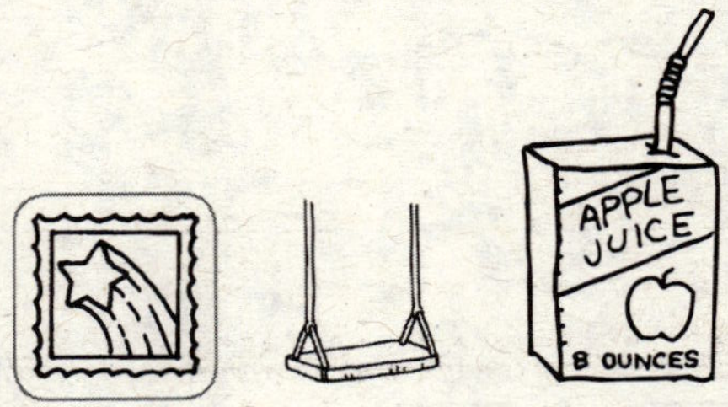

6. Thomas wants to cover the pentagon with 3 shapes. How many of each shape does he need to make the pentagon? I.G.2

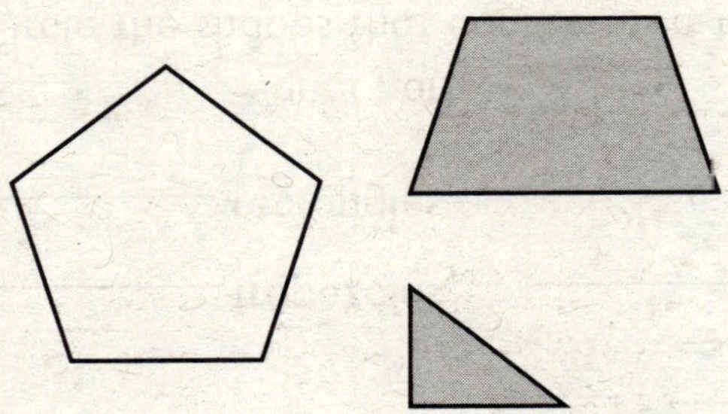

1

2

7. A pizza chef makes a pizza. He cuts it into 4 equal parts. Circle the pictures that could be the pizza. I.G.3

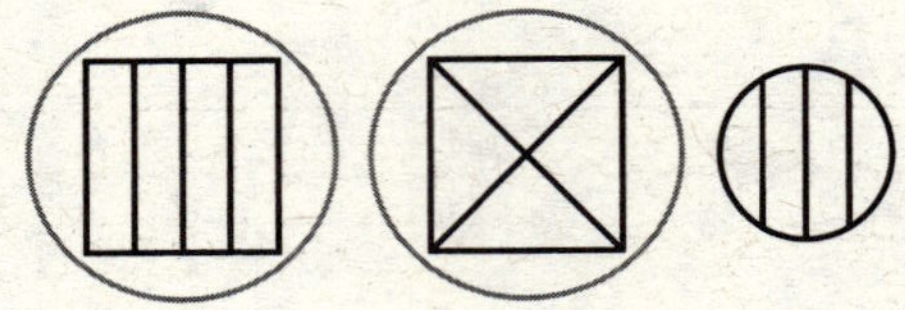

8. True or false?

A trapezoid has 4 vertices and 4 sides. I.G.1

True False

9. Draw a shape that has seven vertices and seven sides. I.G.1

Sample answer is given.

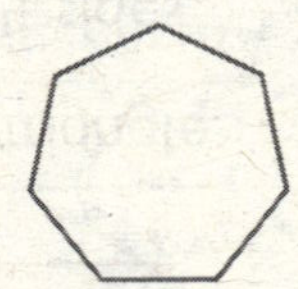

10. Lamar is thinking of a shape with no vertices. What is the shape? I.G.1

A. square

B. circle

C. triangle

11. Enrique is building a table. He splits a wooden board into half. How many equal parts does he have?

2 equal parts

12. Put an X on the shape that could not be used to cover the rectangle. I.G.1

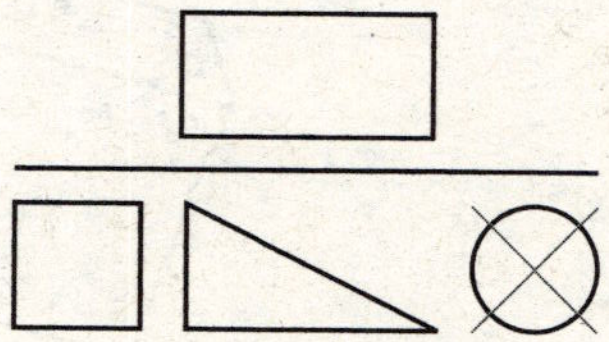

13. Circle the shape that shows thirds. I.G.3

14. Veronica cuts a piece of thread into exact fourths. How many pieces of thread does she have? I.G.3

A. 4

B. 14

C. 40

15. Put an X through any words that do not describe the shape. I.G.1

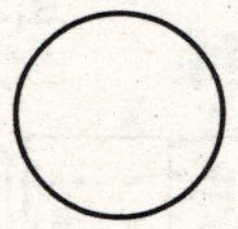

circle

square

0 vertices

3 edges

NAME ________ DATE ________

SCORE ________

Chapter 10 Test

1. Circle all shapes that make up the faces of the sponge. I.G.1

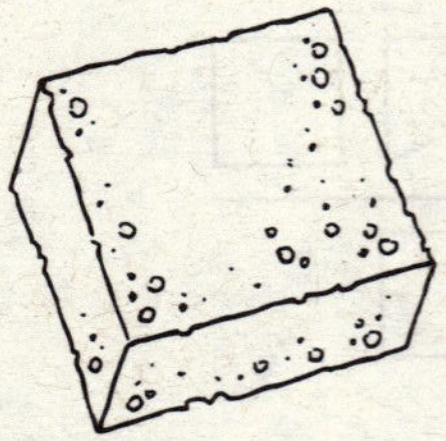

(square) (rectangle)
triangle trapezoid

2. A box of tissues has six square faces. What shape is the box of tissues? I.G.1

cube

3. Circle the shapes that are not used to make the solid. I.G.2

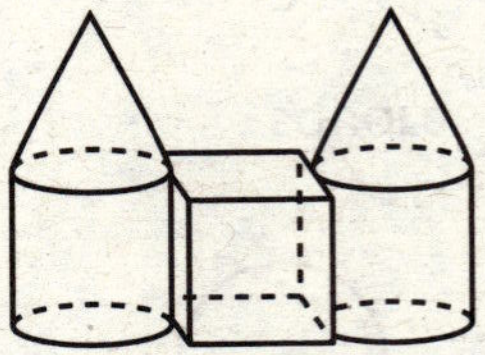

cone (circle) cylinder
cube (triangle)

4. Mr. Craft has a party hat that is shaped like a cone. How many vertices does his hat have? I.G.1

A. 0
(B. 1)
C. 2

5. Juanita made a pattern with blocks. What shape does she need to complete her pattern? I.G.1

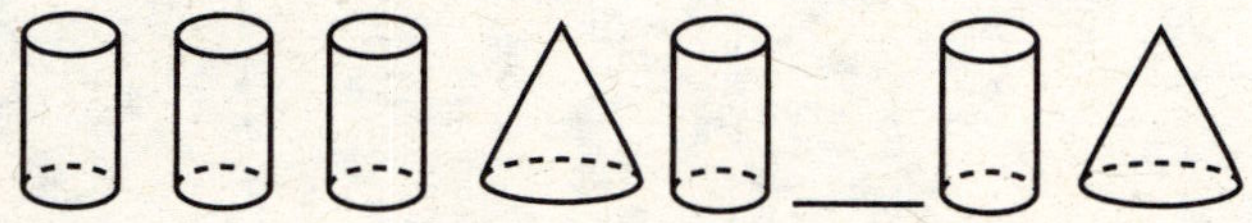

cone — rectangular prism

(cylinder) — cube

6. Underline the shapes that have fewer than six vertices. I.G.1

cone — cylinder

cube — rectangular prism

7. Circle the shape best for packing books in. I.G.1

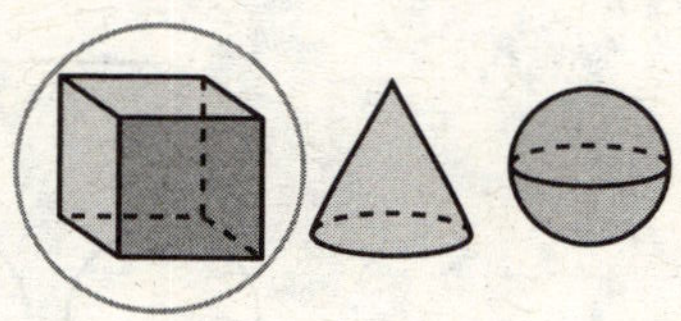

8. A can of tennis balls has 2 faces shaped like circles and no vertices. Circle the name of the shape. I.G.1

sphere — (cylinder) — cone

9. Mrs. Watson is thinking of a shape. All the faces of her shape are the same. What is Mrs. Watson's shape? I.G.1

(A. cube)

B. cone

C. cylinder

10. Circle the two shapes that have the same number of faces. I.G.1

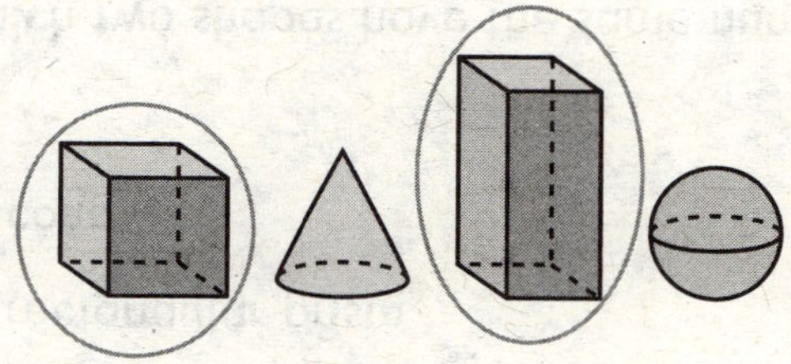

11. The two shapes are faces of a three-dimensional shape. What is the name of the three-dimensional shape? I.G.1

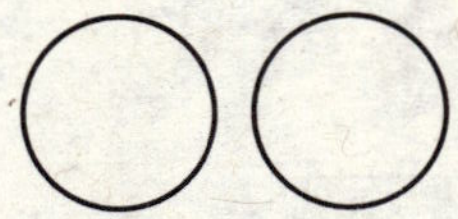

cylinder

12. Sheldon has a collection of blocks that are cubes. He wants to store them in a container. What is the best shape for his container? I.G.2

A. rectangular prism

B. cone

C. cylinder

13. Which shape has a face that is the same shape as the bottom of a cylinder? I.G.1

A. cube

B. rectangular prism

C. cone

14. Which two shapes have the same number of vertices? I.G.1

- ☐ cone
- ☐ cylinder
- ☐ cube
- ☐ rectangular prism

15. Place an X on any shape that will not be in either of the blanks in the pattern. I.G.1

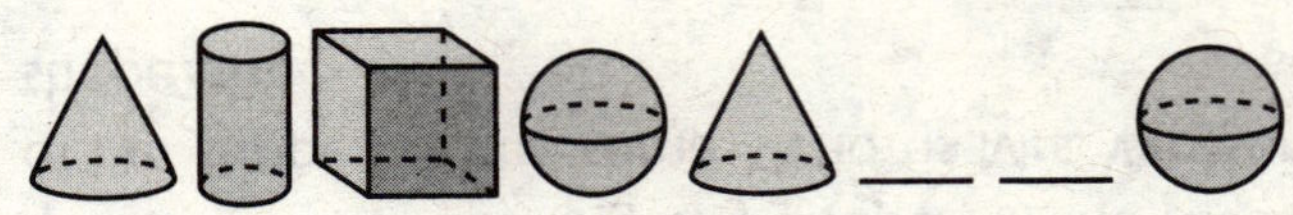

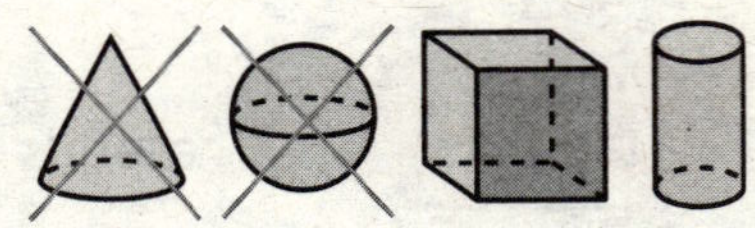

Page 113 • A Lemonade Stand

Task Scenario		
Students will write number sentences, add numbers, and find missing parts to help Felipe and his lemonade stand.		
CCSS Content Standard(s)	1.OA.1, 1.OA.3, 1.OA.6	
Mathematical Practices	MP1, MP2, MP4	
Depth of Knowledge	DOK1, DOK2, DOK3	

Part	Maximum Points	Scoring Rubric
A	2	Full Credit: $4 + 5 = 9$ Partial Credit (1 point) will be given for a correct answer (9) without a correct number sentence. No credit will be given for an incorrect answer.
B	2	Full Credit: $3 + 5 = 8$ Partial Credit (1 point) will be given for a correct answer (8) without a correct number sentence. No credit will be given for an incorrect answer.
C	2	Full Credit: 3 boys No credit will be given for an incorrect answer.
D	2	Full Credit: $7 + 0 = 7$ Partial Credit (1 point) will be given for a correct answer (7) without a correct number sentence. No credit will be given for an incorrect answer.
TOTAL	8	

NAME ____________ DATE ____________

SCORE ____________

Performance Task

A Lemonade Stand

Felipe is selling lemonade. He needs to buy lemons.

Write your answers on another piece of paper. Show all your work to receive full credit.

Part A

1.OA.1

Felipe fills two bags with lemons. Write an addition sentence to show the number of lemons in all.

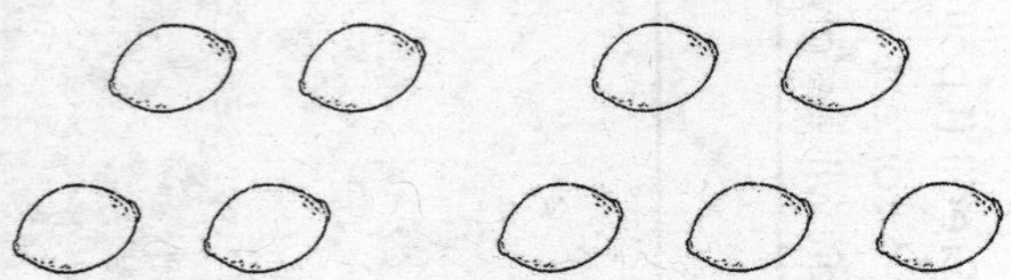

4 + 5 = 9 lemons

Part B

1.OA.1

Felipe sells 3 cups of lemonade in the morning and 5 cups in the afternoon. How many cups did Felipe sell in all?

3 + 5 = 8 cups

Performance Task *(continued)*

Part C

1.OA.6

10 people waved to Felipe. 7 were girls. Find the missing part to find out how many were boys.

Part	Part
7 girls	3 boys
Whole	
10 people	

Part D

1.OA.3

Felipe also gave away water for free. In the morning he gave away 7 cups of water. In the afternoon he gave away 0 cups of water. How many cups did he give away in all?

7 + 0 = 7 cups

NAME ____________ DATE ____________

SCORE ____________

Performance Task

A Lemonade Stand

Felipe is selling lemonade. He needs to buy lemons.

Write your answers on another piece of paper. Show all your work to receive full credit.

Part A

I.OA.1

Felipe fills two bags with lemons. Write an addition sentence to show the number of lemons in all.

4 + 4 = 8 lemons

Part B

I.OA.1

Felipe sells 3 cups of lemonade in the morning and 5 cups in the afternoon. How many cups did Felipe sell in all?

3 + 5 = 9 cups

Performance Task *(continued)*

Part C

I.OA.6

10 people waved to Felipe. 7 were girls. Find the missing part to find out how many were boys.

Part	Part
7 girls	3 boys
Whole	
10 people	

Part D

I.OA.3

Felipe also gave away water for free. In the morning he gave away 7 cups of water. In the afternoon he gave away 0 cups of water. How many cups did he give away in all?

7 + 0 = 7 cups

Copyright © McGraw-Hill Education. Permission is granted to reproduce for classroom use.

Student Model

NAME ______________ DATE ______________

SCORE ______________

Performance Task

A Lemonade Stand

Felipe is selling lemonade. He needs to buy lemons.

Write your answers on another piece of paper. Show all your work to receive full credit.

Part A

1.OA.1

Felipe fills two bags with lemons. Write an addition sentence to show the number of lemons in all.

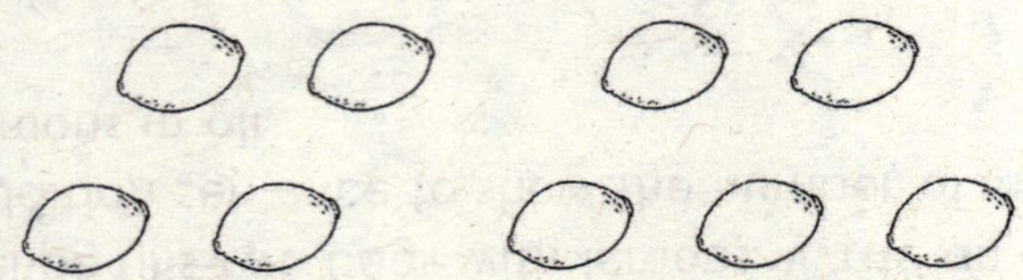

____ + ____ = 9 lemons

Part B

1.OA.1

Felipe sells 3 cups of lemonade in the morning and 5 cups in the afternoon. How many cups did Felipe sell in all?

____ + ____ = 8 cups

Performance Task *(continued)*

Part C

1.OA.6

10 people waved to Felipe. 7 were girls. Find the missing part to find out how many were boys.

Part	Part
7 girls	3 boys
Whole	
10 people	

Part D

1.OA.3

Felipe also gave away water for free. In the morning he gave away 7 cups of water. In the afternoon he gave away 0 cups of water. How many cups did he give away in all?

____ + ____ = 7 cups

NAME ____________ DATE ____________

SCORE ____________

Performance Task

A Lemonade Stand

Felipe is selling lemonade. He needs to buy lemons.

Write your answers on another piece of paper. Show all your work to receive full credit.

Part A

I.OA.I

Felipe fills two bags with lemons. Write an addition sentence to show the number of lemons in all.

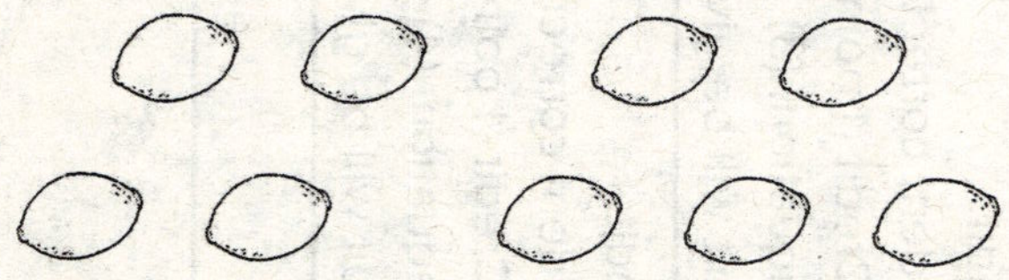

____ + ____ = 8 lemons

Part B

I.OA.I

Felipe sells 3 cups of lemonade in the morning and 5 cups in the afternoon. How many cups did Felipe sell in all?

____ + ____ = 9 cups

Performance Task *(continued)*

Part C

I.OA.6

10 people waved to Felipe. 7 were girls. Find the missing part to find out how many were boys.

Part	Part
7 girls	2 boys
Whole	
10 people	

Part D

I.OA.3

Felipe also gave away water for free. In the morning he gave away 7 cups of water. In the afternoon he gave away 0 cups of water. How many cups did he give away in all?

____ + ____ = 7 cups

Student Model

Page 115 • At the Library

Task Scenario Students will solve subtraction word problems, write subtraction equations, and identify related subtraction facts.		
CCSS Content Standard(s)		1.OA.1, 1.OA.6
Mathematical Practices		MP1, MP2, MP4
Depth of Knowledge		DOK1, DOK2, DOK3
Part	**Maximum Points**	**Scoring Rubric**
A	2	Full Credit: The student correctly writes the equation. (8 − 8 = 0) Partial Credit (1 point) will be given for giving the correct solution to the equation. (0) No credit will be given for an incorrect answer.
B	2	Full Credit: The student correctly writes the equation. (8 − 5 = 3) Partial Credit (1 point) will be given for giving the correct solution to the equation. (3) No credit will be given for an incorrect answer.
C	2	Full Credit: The student correctly writes the equation. (5 − 3 = 2) Partial Credit (1 point) will be given for giving the correct solution to the equation. (2) No credit will be given for an incorrect answer.
D	2	Full Credit: The student correctly writes the equation. (5 − 2 = 3) Partial Credit (1 point) will be given for giving the correct solution to the equation. (3) No credit will be given for an incorrect answer.
E	2	Full Credit: The student correctly writes the equation. (7 − 3 = 4) Partial Credit (1 point) will be given for giving the correct solution to the equation. (4) No credit will be given for an incorrect answer.
TOTAL	10	

NAME ____________ DATE ____________

SCORE ____________

Performance Task

At the Library

Mr. Fowler works at the library. He checks out books for people.

Part A

Mr. Fowler's first job of the day is checking in books that have been returned. Write a subtraction number sentence to show what he did.

8 – 8 = 0

Part B

Mr. Fowler puts the 8 books on his cart. He takes it to the shelf and puts away 5 books. How many books are left on the shelf?

8 – 5 = 3

Performance Task *(continued)*

Part C

Mr. Fowler has a line of 5 people to help. He helps 3 people. How many people are left?

5 – 3 = 2 people

Part D

Write the related subtraction sentence to show how many people Mr. Fowler helped.

5 – 2 = 3

Part E

Mr. Fowler had 7 story time books on a shelf. The boys and girls who came to the library borrowed 3 books. Write the subtraction number sentence.

7 – 3 = 4 books

Student Model

NAME ________ DATE ________

SCORE ________

Performance Task

At the Library

Mr. Fowler works at the library. He checks out books for people.

Part A

Mr. Fowler's first job of the day is checking in books that have been returned. Write a subtraction number sentence to show what he did.

____ − ____ = __0__

Part B

Mr. Fowler puts the 8 books on his cart. He takes it to the shelf and puts away 5 books. How many books are left on the shelf?

__8__ − __5__ = __3__

Performance Task *(continued)*

Part C

Mr. Fowler has a line of 5 people to help. He helps 3 people. How many people are left?

__5__ − __3__ = __2__ people

Part D

Write the related subtraction sentence to show how many people Mr. Fowler helped.

____ − ____ = __3__

Part E

Mr. Fowler had 7 story time books on a shelf. The boys and girls who came to the library borrowed 3 books. Write the subtraction number sentence.

__7__ − __3__ = __4__ books

NAME ________ DATE ________

SCORE ________

Performance Task

At the Library

Mr. Fowler works at the library. He checks out books for people.

Part A

Mr. Fowler's first job of the day is checking in books that have been returned. Write a subtraction number sentence to show what he did.

___ – ___ = 0

Part B

Mr. Fowler puts the 8 books on his cart. He takes it to the shelf and puts away 5 books. How many books are left on the shelf?

___ – ___ = 3

Performance Task *(continued)*

Part C

Mr. Fowler has a line of 5 people to help. He helps 3 people. How many people are left?

___ – ___ = 2 people

Part D

Write the related subtraction sentence to show how many people Mr. Fowler helped.

___ – ___ = 3

Part E

Mr. Fowler had 7 story time books on a shelf. The boys and girls who came to the library borrowed 3 books. Write the subtraction number sentence.

___ – ___ = 4 books

Student Model

NAME .... DATE ....

SCORE ....

Performance Task

At the Library

Mr. Fowler works at the library. He checks out books for people.

Part A

Mr. Fowler's first job of the day is checking in books that have been returned. Write a subtraction number sentence to show what he did.

_____ – _____ = 0

Part B

Mr. Fowler puts the 8 books on his cart. He takes it to the shelf and puts away 5 books. How many books are left on the shelf?

8 – 2 = 4

Performance Task *(continued)*

Part C

Mr. Fowler has a line of 5 people to help. He helps 3 people. How many people are left?

5 – 4 = 0 people

Part D

Write the related subtraction sentence to show how many people Mr. Fowler helped.

_____ – _____ = 3

Part E

Mr. Fowler had 7 story time books on a shelf. The boys and girls who came to the library borrowed 3 books. Write the subtraction number sentence.

_____ – _____ = 7 books

Page 117 • Picking Apples

Task Scenario		
Students will use number lines and "making ten" to add two or three numbers together.		
CCSS Content Standard(s)		1.OA.2, 1.OA.3, 1.OA.6
Mathematical Practices		MP1, MP2, MP4
Depth of Knowledge		DOK1, DOK2, DOK3
Part	**Maximum Points**	**Scoring Rubric**
A	3	Full Credit: 16 apples 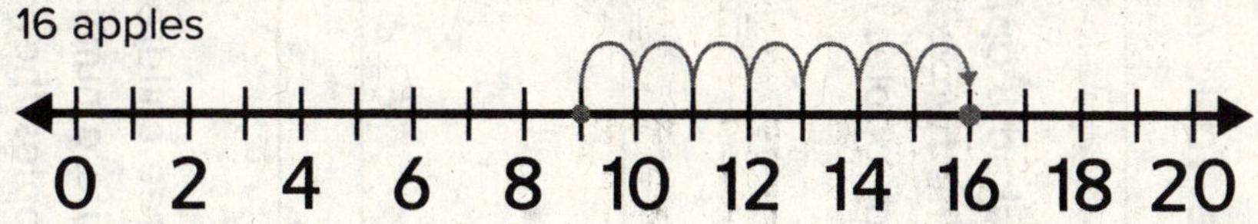 Partial Credit (1 point) will be given for a correct answer without a number line. Another point will be given for having the correct dots on the number line without the arches showing addition. No credit will be given for an incorrect answer.
B	3	Full Credit: 8 + 4 = 12 apples 4 + 8 = 12 apples Partial Credit (1 point) will be given for a correct answer (12). An additional point (1 point) will be given for having at least one correct number sentence. No credit will be given for an incorrect answer.
C	2	Full Credit: 5 + 4 + 6 = 15 apples Partial Credit (1 point) will be given for a correct answer without a correct number sentence. No credit will be given for an incorrect answer.
TOTAL	8	

NAME ____________ DATE ____________

SCORE ____________

Performance Task

Picking Apples

The Perez family is picking apples.

Write your answers on another piece of paper. Show all your work to receive full credit.

Part A

I.OA.6

Miguel picked 9 apples. His sister Juanita picked 7 apples. Use the number line to show how many apples they picked in all.

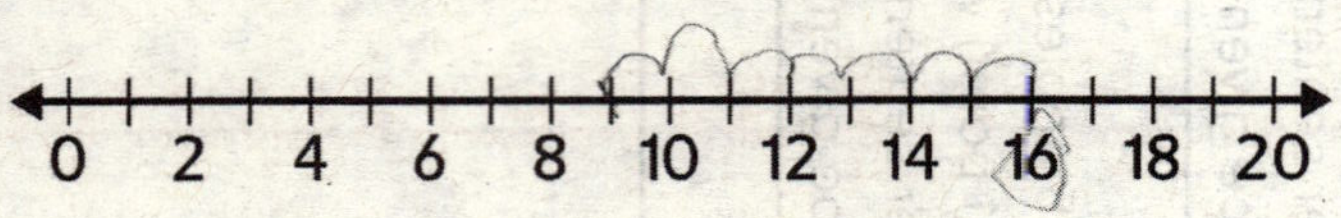

$$\begin{array}{r} 9 \\ +\,7 \\ \hline 16 \end{array}$$ apples

Performance Task *(continued)*

Part B

I.OA.3

Mrs. Perez picked 8 apples. Mr. Perez picked 4 apples. Write two addition sentences to find how many apples they picked in all.

8 + 4 = 12 apples

4 + 8 = 12 apples

Part C

I.OA.2

The other three children picked 5, 4, and 6 apples. Write a number sentence to find how many apples they picked in all.

5 + 4 + 6 = 15 apples

NAME ____________ DATE ____________

SCORE ____________

Performance Task

Picking Apples

The Perez family is picking apples.

Write your answers on another piece of paper. Show all your work to receive full credit.

Part A

I.OA.6

Miguel picked 9 apples. His sister Juanita picked 7 apples. Use the number line to show how many apples they picked in all.

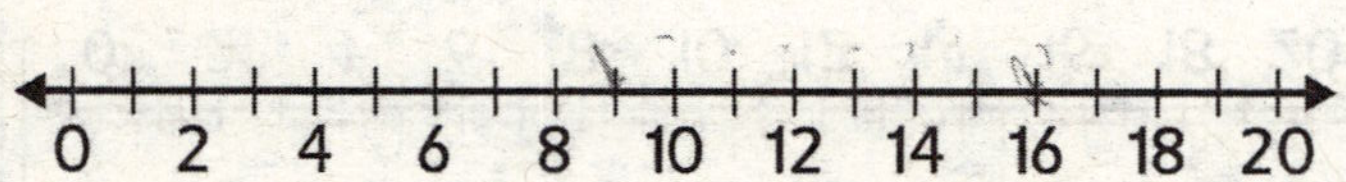

$$\begin{array}{r} 9 \\ +\ 7 \\ \hline 16 \end{array}$$ apples

Performance Task *(continued)*

Part B

I.OA.3

Mrs. Perez picked 8 apples. Mr. Perez picked 4 apples. Write two addition sentences to find how many apples they picked in all.

8 + 4 = 12 apples

___ + ___ = 12 apples

Part C

I.OA.2

The other three children picked 5, 4, and 6 apples. Write a number sentence to find how many apples they picked in all.

5 + 4 + 6 = 15 apples

Student Model

NAME ____________ DATE ____________

SCORE ____________

Performance Task

Picking Apples

The Perez family is picking apples.

Write your answers on another piece of paper. Show all your work to receive full credit.

Part A

I.OA.6

Miguel picked 9 apples. His sister Juanita picked 7 apples. Use the number line to show how many apples they picked in all.

  9
+ 7
16 apples

Performance Task *(continued)*

Part B

I.OA.3

Mrs. Perez picked 8 apples. Mr. Perez picked 4 apples. Write two addition sentences to find how many apples they picked in all.

____ + ____ = 8 apples

____ + ____ = 4 apples

Part C

I.OA.2

The other three children picked 5, 4, and 6 apples. Write a number sentence to find how many apples they picked in all.

5 + 4 + 6 = 14 apples

NAME ________________ DATE ________________

SCORE ________________

Performance Task

Picking Apples

The Perez family is picking apples.

Write your answers on another piece of paper. Show all your work to receive full credit.

Part A

I.OA.6

Miguel picked 9 apples. His sister Juanita picked 7 apples. Use the number line to show how many apples they picked in all.

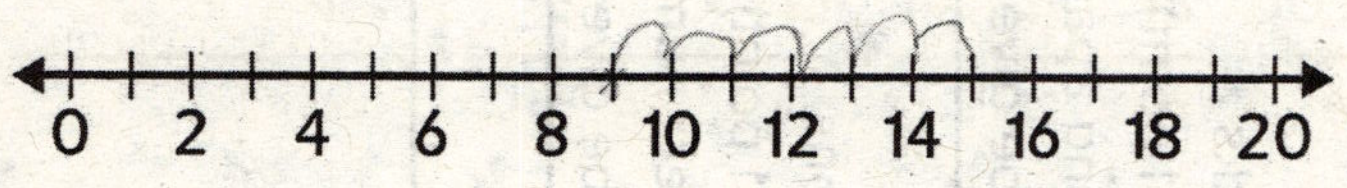

$$\begin{array}{r} 9 \\ +\ 7 \\ \hline 15 \end{array}$$ apples

Performance Task *(continued)*

Part B

I.OA.3

Mrs. Perez picked 8 apples. Mr. Perez picked 4 apples. Write two addition sentences to find how many apples they picked in all.

_____ + _____ = 12 apples

_____ + _____ = _____ apples

Part C

I.OA.2

The other three children picked 5, 4, and 6 apples. Write a number sentence to find how many apples they picked in all.

4 + 5 + 6 = 13 apples

Student Model

Page 119 • Animals at the Zoo

Task Scenario		
Students will use subtraction strategies for numbers up to 20 to find how many animals are left.		
CCSS Content Standard(s)		1.OA.4, 1.OA.5, 1.OA.6
Mathematical Practices		MP1, MP2, MP4
Depth of Knowledge		DOK1, DOK2, DOK3, DOK4
Part	**Maximum Points**	**Scoring Rubric**
A	3	Full Credit:  13 – 5 = 8 monkeys Partial Credit (1 point) will be given for a correct number line. An additional point (1 point) will be given for a correct answer (8) with an incorrect number sentence. No credit will be given for an incorrect answer.
B	1	Full Credit: 8 + 5 = 13 monkeys No credit will be given for an incorrect number sentence.
C	2	Full Credit: 4 = 1 + 3 11 – 4 = 7 ducks Partial Credit (1 point) will be given for a correct answer (7) without breaking apart 4 correctly into 1 and 3. No credit will be given for an incorrect answer.
D	2	Full Credit: 15 – 5 = 10 lions Partial Credit (1 point) will be given for a correct answer (10) without a correct number sentence. No credit will be given for an incorrect answer.
TOTAL	8	

NAME ______________ DATE ______________

SCORE ______________

Performance Task

Animals at the Zoo

Kerin takes a trip to the zoo with his class. He sees a lot of animals.

Write your answers on another piece of paper. Show all your work to receive full credit.

Part A

I.OA.5

Kerin counted 13 monkeys in a tree. 5 of the monkeys climb down. Use the number line to find how many monkeys are still in the tree.

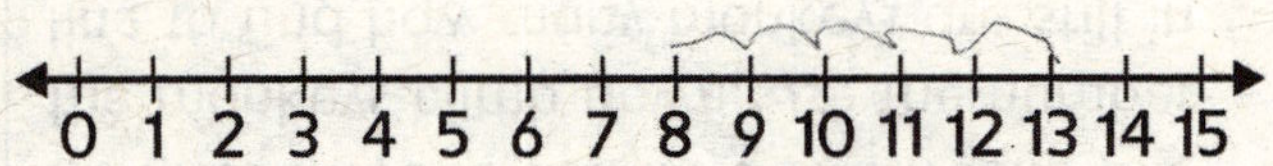

13 – 5 = 8 monkeys left

Performance Task *(continued)*

Part B

I.OA.6

Write a related addition sentence for the monkeys Kerin saw.

8 + 5 = 13 monkeys

Part C

I.OA.6

Kerin saw 11 ducks swimming in the pond. 4 ducks flew out of the water. How many ducks were left? Take apart the number to make 10. Then subtract.

11 – 4

1 3

11 – 4 = 7 ducks left

Part D

I.OA.4

Kerin saw 15 lions. 5 were sleeping. How many lions were awake? Write a number sentence.

15 – 5 = 10 lions

Student Model

NAME ______________ DATE ______________

SCORE ______________

Performance Task

Animals at the Zoo

Kerin takes a trip to the zoo with his class. He sees a lot of animals.

Write your answers on another piece of paper. Show all your work to receive full credit.

Part A

I.OA.5

Kerin counted 13 monkeys in a tree. 5 of the monkeys climb down. Use the number line to find how many monkeys are still in the tree.

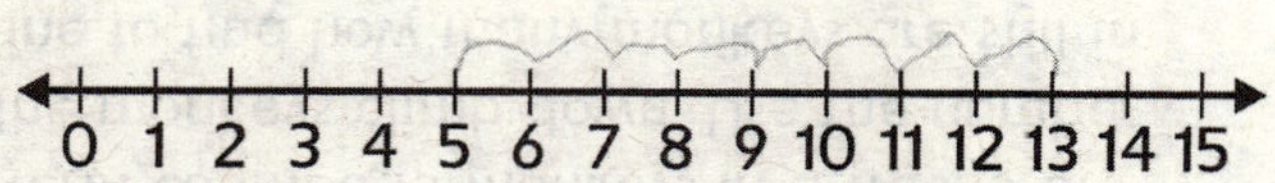

13 − 8 = 5 monkeys left

Performance Task *(continued)*

Part B

I.OA.6

Write a related addition sentence for the monkeys Kerin saw.

8 + 5 = 13 monkeys

Part C

I.OA.6

Kerin saw 11 ducks swimming in the pond. 4 ducks flew out of the water. How many ducks were left? Take apart the number to make 10. Then subtract.

11 − 4

[2] [2]

11 − 4 = 7 ducks left

Part D

I.OA.4

Kerin saw 15 lions. 5 were sleeping. How many lions were awake? Write a number sentence.

_____ − _____ = 10 lions

NAME ____________ DATE ____________

SCORE ____________

Performance Task

Animals at the Zoo

Kerin takes a trip to the zoo with his class. He sees a lot of animals.

Write your answers on another piece of paper. Show all your work to receive full credit.

Part A 1.OA.5

Kerin counted 13 monkeys in a tree. 5 of the monkeys climb down. Use the number line to find how many monkeys are still in the tree.

0 1 2 3 4 5 6 7 8 9 10 11 12 13 14 15

13 – 8 = 5 monkeys left

Performance Task *(continued)*

Part B 1.OA.6

Write a related addition sentence for the monkeys Kerin saw.

8 + 5 = 13 monkeys

Part C 1.OA.6

Kerin saw 11 ducks swimming in the pond. 4 ducks flew out of the water. How many ducks were left? Take apart the number to make 10. Then subtract.

11 – 4

11 – 4 = 6 ducks left

Part D 1.OA.4

Kerin saw 15 lions. 5 were sleeping. How many lions were awake? Write a number sentence.

15 – 5 = 9 lions

Student Model

NAME ____________ DATE ____________

SCORE ____________

Performance Task

Animals at the Zoo

Kerin takes a trip to the zoo with his class. He sees a lot of animals.

Write your answers on another piece of paper. Show all your work to receive full credit.

Part A

I.OA.5

Kerin counted 13 monkeys in a tree. 5 of the monkeys climb down. Use the number line to find how many monkeys are still in the tree.

13 – 5 = 7 monkeys left

Performance Task *(continued)*

Part B

I.OA.6

Write a related addition sentence for the monkeys Kerin saw.

7 + 5 = 13 monkeys

Part C

I.OA.6

Kerin saw 11 ducks swimming in the pond. 4 ducks flew out of the water. How many ducks were left? Take apart the number to make 10. Then subtract.

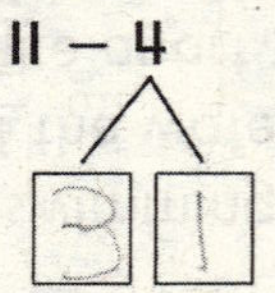

11 – 4 = 6 ducks left

Part D

I.OA.4

Kerin saw 15 lions. 5 were sleeping. How many lions were awake? Write a number sentence.

5 + 10 = 15 lions

Page 121 • Saving Money

Task Scenario Students will use dimes, nickels, and pennies to count money and compare numbers.		
CCSS Content Standard(s)		1.OA.6, 1.NBT.1, 1.NBT.3
Mathematical Practices		MP1, MP2, MP4
Depth of Knowledge		DOK2, DOK3
Part	**Maximum Points**	**Scoring Rubric**
A	2	Full Credit: 8¢ + 5¢ = 13¢ Partial Credit (1 point) will be given a correct answer without a correct number sentence. No credit will be given for an incorrect answer.
B	2	Full Credit: 10¢, 20¢, 30¢, 40¢, 50¢, 60¢ Partial Credit (1 point) will be given for starting wrong but still counting by 10s (20¢, 30¢, 40¢, ...). No credit will be given for not counting by 10s correctly.
C	2	Full Credit: 17 nickels No credit will be given for an incorrect answer.
D	2	Full Credit: 36 < 63 Jack Partial Credit (1 point) will be given for having the correct sign **OR** the correct name. No credit will be given for two incorrect answers.
TOTAL	8	

Performance Task Rubrics

NAME ______________________ DATE ______________

SCORE ______________

Performance Task

Saving Money

Bree is saving money to buy a toy. She has nickels, pennies, and dimes.

Write your answers on another piece of paper. Show all your work to receive full credit.

Part A

I.OA.6

Bree has 8 pennies in her piggy bank. She finds 5 pennies on the ground. How much money does Bree have in all?

8 ¢ + 5 ¢ = 13 ¢

Performance Task *(continued)*

Part B

I.NBT.1

Bree takes the dimes out of her piggy bank. How much money does she have in dimes?

10 ¢ 20 ¢ 30 ¢ 40 ¢ 50 ¢ 60 ¢ in all

Part C

I.NBT.1, I.NBT.3

Bree takes out her nickels. She has 85¢ in nickels. How many nickels does Bree have?

17 nickels

Part D

I.NBT.3

Bree has 36 coins in all. Her brother Jack has 63 coins. Write < or >, and tell who has more coins.

36 (<) 63

Who has more? Jack

NAME ________ DATE ________

SCORE ________

Performance Task

Saving Money

Bree is saving money to buy a toy. She has nickels, pennies, and dimes.

Write your answers on another piece of paper. Show all your work to receive full credit.

Part A

I.OA.6

Bree has 8 pennies in her piggy bank. She finds 5 pennies on the ground. How much money does Bree have in all?

8 ¢ + 5 ¢ = 12 ¢

Performance Task *(continued)*

Part B

I.NBT.1

Bree takes the dimes out of her piggy bank. How much money does she have in dimes?

10 ¢ 20 ¢ 30 ¢ 40 ¢ 50 ¢ 60 ¢ in all

Part C

I.NBT.1, I.NBT.3

Bree takes out her nickels. She has 85¢ in nickels. How many nickels does Bree have?

15 nickels

Part D

I.NBT.3

Bree has 36 coins in all. Her brother Jack has 63 coins. Write < or >, and tell who has more coins.

36 (<) 63

Who has more? Jack

Student Model

NAME ________________ DATE ________

SCORE ________

Performance Task

Saving Money

Bree is saving money to buy a toy. She has nickels, pennies, and dimes.

Write your answers on another piece of paper. Show all your work to receive full credit.

Part A I.OA.6

Bree has 8 pennies in her piggy bank. She finds 5 pennies on the ground. How much money does Bree have in all?

____¢ + ____¢ = 13¢

Performance Task *(continued)*

Part B I.NBT.1

Bree takes the dimes out of her piggy bank. How much money does she have in dimes?

____¢ ____¢ ____¢ ____¢ ____¢ 60¢ in all

Part C I.NBT.1, I.NBT.3

Bree takes out her nickels. She has 85¢ in nickels. How many nickels does Bree have?

19 nickels

Part D I.NBT.3

Bree has 36 coins in all. Her brother Jack has 63 coins. Write < or >, and tell who has more coins.

36 (<) 63

Who has more? ________

NAME ________________ DATE ________

SCORE ________

Performance Task

Saving Money

Bree is saving money to buy a toy. She has nickels, pennies, and dimes.

Write your answers on another piece of paper. Show all your work to receive full credit.

Part A

I.OA.6

Bree has 8 pennies in her piggy bank. She finds 5 pennies on the ground. How much money does Bree have in all?

8 ¢ + 5 ¢ = 14 ¢

Performance Task *(continued)*

Part B

I.NBT.1

Bree takes the dimes out of her piggy bank. How much money does she have in dimes?

10 ¢ 10 ¢ 10 ¢ 10 ¢ 10 ¢ 10 ¢ in all

Part C

I.NBT.1, I.NBT.3

Bree takes out her nickels. She has 85¢ in nickels. How many nickels does Bree have?

15 nickels

Part D

I.NBT.3

Bree has 36 coins in all. Her brother Jack has 63 coins. Write < or >, and tell who has more coins.

36 (>) 63

Who has more? Bree

Student Model

Page 123 • Buying School Supplies

Task Scenario Students will use two-digit addition and subtraction to find how many school supplies there are.		
CCSS Content Standard(s)		1.NBT.4, 1.NBT.6
Mathematical Practices		MP1, MP2, MP4
Depth of Knowledge		DOK2, DOK3
Part	**Maximum Points**	**Scoring Rubric**
A	2	Full Credit: 30 + 40 = 70 pens Partial Credit (1 point) will be given for a correct answer without a correct number sentence. No credit will be given for an incorrect answer.
B	2	Full Credit: 24 + 30 = 54 notebooks Partial Credit (1 point) will be given for a correct answer without a correct number sentence. No credit will be given for an incorrect answer.
C	3	Full Credit: 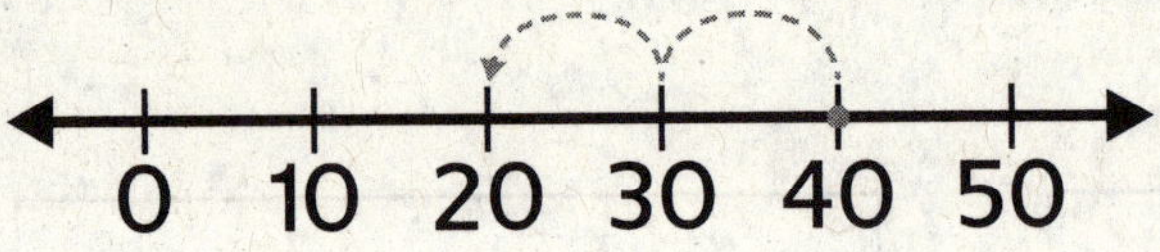 40 − 20 = 20 boxes Partial Credit (1 point) will be given for a correct answer (20). Another point will be given for a correct number line **OR** a correct number sentence. No credit will be given for an incorrect answer.
D	2	Full Credit: 8 + 31 = 39 pencils Partial Credit (1 point) will be given for a correct answer without a correct number sentence. No credit will be given for an incorrect answer.
TOTAL	9	

NAME ________________ DATE ________

SCORE ________

Performance Task

Buying School Supplies

Ms. Chatlos is buying supplies for her 1st grade class.

Write your answers on another piece of paper. Show all your work to receive full credit.

Part A I.NBT.4

Ms. Chatlos buys 30 blue pens and 40 black pens. Write a number sentence to find how many pens she buys in all.

30 + 40 = 70 pens

Part B I.NBT.4

Ms. Chatlos buys 24 small notebooks and 30 big notebooks. Write a number sentence to find how many notebooks she buys in all.

24 + 30 = 54 notebooks

Performance Task *(continued)*

Part C I.NBT.6

Ms. Chatlos buys 40 boxes of crayons. She returns 20 boxes. How many boxes does she have left? Use the number line.

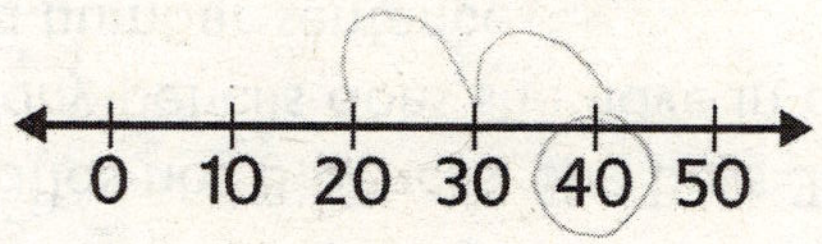

40 – 20 = 20 boxes

Part D I.NBT.4

Ms. Chatlos has 8 pencils. She buys 31 more. How many pencils does she have in all? Write a number sentence.

8 + 31 = 39 pencils

Student Model

NAME ________ DATE ________

SCORE ________

Performance Task

Buying School Supplies

Ms. Chatlos is buying supplies for her 1st grade class.

Write your answers on another piece of paper. Show all your work to receive full credit.

Part A 1.NBT.4

Ms. Chatlos buys 30 blue pens and 40 black pens. Write a number sentence to find how many pens she buys in all.

30 + 40 = 70 pens

Part B 1.NBT.4

Ms. Chatlos buys 24 small notebooks and 30 big notebooks. Write a number sentence to find how many notebooks she buys in all.

20 + 30 = 50 notebooks

Performance Task *(continued)*

Part C 1.NBT.6

Ms. Chatlos buys 40 boxes of crayons. She returns 20 boxes. How many boxes does she have left? Use the number line.

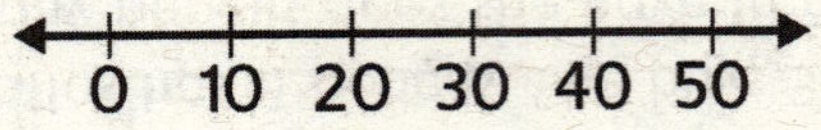

40 – 20 = 20 boxes

Part D 1.NBT.4

Ms. Chatlos has 8 pencils. She buys 31 more. How many pencils does she have in all? Write a number sentence.

8 + 30 = 38 pencils

NAME ______________ DATE ______________

SCORE ______________

Performance Task

Buying School Supplies

Ms. Chatlos is buying supplies for her 1st grade class.

Write your answers on another piece of paper. Show all your work to receive full credit.

Part A 1.NBT.4

Ms. Chatlos buys 30 blue pens and 40 black pens. Write a number sentence to find how many pens she buys in all.

30 + 4 = 34 pens

Part B 1.NBT.4

Ms. Chatlos buys 24 small notebooks and 30 big notebooks. Write a number sentence to find how many notebooks she buys in all.

_____ + _____ = 54 notebooks

Performance Task *(continued)*

Part C 1.NBT.6

Ms. Chatlos buys 40 boxes of crayons. She returns 20 boxes. How many boxes does she have left? Use the number line.

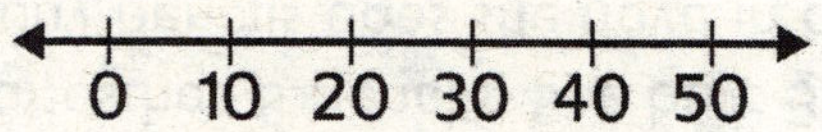

_____ − _____ = 20 boxes

Part D 1.NBT.4

Ms. Chatlos has 8 pencils. She buys 31 more. How many pencils does she have in all? Write a number sentence.

31 + 9 = 40 pencils

Student Model

NAME ____________ DATE ____________

SCORE ____________

Performance Task

Buying School Supplies

Ms. Chatlos is buying supplies for her 1st grade class.

Write your answers on another piece of paper. Show all your work to receive full credit.

Part A 1.NBT.4

Ms. Chatlos buys 30 blue pens and 40 black pens. Write a number sentence to find how many pens she buys in all.

30 + 40 = 60 pens

Part B 1.NBT.4

Ms. Chatlos buys 24 small notebooks and 30 big notebooks. Write a number sentence to find how many notebooks she buys in all.

30 + ____ = 54 notebooks

Performance Task *(continued)*

Part C 1.NBT.6

Ms. Chatlos buys 40 boxes of crayons. She returns 20 boxes. How many boxes does she have left? Use the number line.

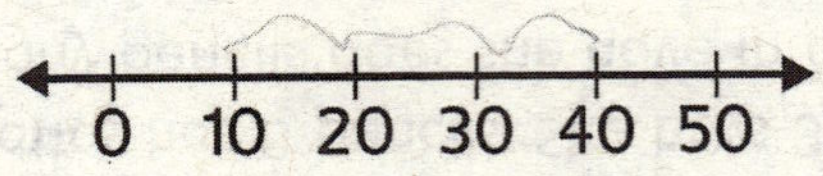

40 – 20 = 10 boxes

Part D 1.NBT.4

Ms. Chatlos has 8 pencils. She buys 31 more. How many pencils does she have in all? Write a number sentence.

____ + ____ = 40 pencils

Page 125 • Favorite Sport

<table>
<tr><td colspan="3">Task Scenario
Students will use bar graphs and tally charts to answer questions about people's favorite sport.</td></tr>
<tr><td colspan="2">CCSS Content Standard(s)</td><td>1.NBT.4, 1.MD.4</td></tr>
<tr><td colspan="2">Mathematical Practices</td><td>MP1, MP2, MP4, MP5</td></tr>
<tr><td colspan="2">Depth of Knowledge</td><td>DOK2, DOK3, DOK4</td></tr>
<tr><th>Part</th><th>Maximum Points</th><th>Scoring Rubric</th></tr>
<tr><td>A</td><td>3</td><td>Full Credit:
<table>
<tr><th colspan="3">Favorite Sports</th></tr>
<tr><th>Sport</th><th>Tally</th><th>Total</th></tr>
<tr><td>Soccer</td><td>卌 卌 |</td><td>11</td></tr>
<tr><td>Football</td><td>||</td><td>2</td></tr>
<tr><td>Baseball</td><td>卌 |</td><td>6</td></tr>
</table>
Partial Credit (1 point) will be given for each correct chart entry (11, 2, and 6).
No credit will be given for three incorrect chart entries.</td></tr>
<tr><td>B</td><td>1</td><td>Full Credit:
football
No credit will be given for an incorrect answer.</td></tr>
<tr><td>C</td><td>3</td><td>Full Credit:

Partial Credit (1 point) will be given for each correct chart entry (11, 2, and 6).
No credit will be given for three incorrect chart entries.</td></tr>
</table>

Part	Maximum Points	Scoring Rubric
D	2	Full Credit: $11 + 2 + 6 = 19$ students Partial Credit (1 point) will be given for a correct answer without a correct number sentence. No credit will be given for an incorrect answer.
TOTAL	9	

NAME ______________ DATE ______________

SCORE ______________

Performance Task

Favorite Sport

Jude asked the students in his class their favorite sport. The table shows how many students chose each sport.

Favorite Sport	
Soccer	11
Football	2
Baseball	6

Write your answers on another piece of paper. Show all your work to receive full credit.

Part A

I.MD.4

Write the tallies for each sport.

Favorite Sports		
Sport	Tally	Total
Soccer	|||| |||| |	11
Football	||	2
Baseball	|||| |	6

Performance Task *(continued)*

Part B

I.MD.4

What sport do people like the least?

Answer: football

Part C

I.MD.4

Make a bar graph.

Part D

I.MD.4, I.NBT.4

How many students are in Jude's class? Write a number sentence.

11+2+6=19

Student Model

NAME ______________ DATE ______________

SCORE ______________

Performance Task

Favorite Sport

Jude asked the students in his class their favorite sport. The table shows how many students chose each sport.

Favorite Sport	
Soccer	11
Football	2
Baseball	6

Write your answers on another piece of paper. Show all your work to receive full credit.

Part A I.MD.4

Write the tallies for each sport.

Favorite Sports		
Sport	Tally	Total
Soccer		11
Football		2
Baseball		6

Performance Task *(continued)*

Part B I.MD.4

What sport do people like the least?

Answer: Football

Part C I.MD.4

Make a bar graph.

Part D I.MD.4, I.NBT.4

How many students are in Jude's class?
Write a number sentence.

19

NAME ______________________ DATE ______________

SCORE ______________

Performance Task

Favorite Sport

Jude asked the students in his class their favorite sport. The table shows how many students chose each sport.

Favorite Sport	
Soccer	11
Football	2
Baseball	6

Write your answers on another piece of paper. Show all your work to receive full credit.

Part A

I.MD.4

Write the tallies for each sport.

Favorite Sports		
Sport	Tally	Total
Soccer		11
Football		2
Baseball		6

Performance Task *(continued)*

Part B

I.MD.4

What sport do people like the least?

Answer: Soccer

Part C

I.MD.4

Make a bar graph.

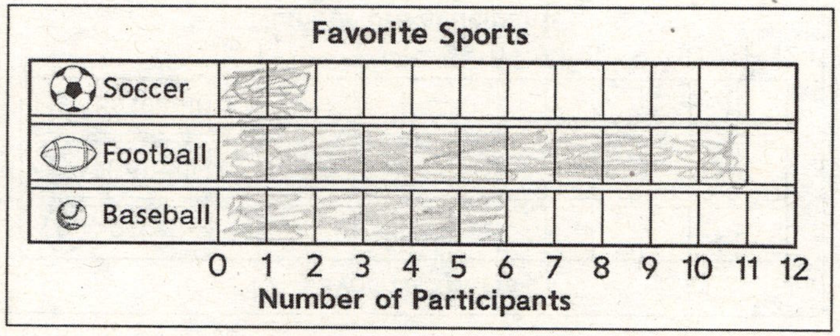

Part D

I.MD.4, I.NBT.4

How many students are in Jude's class?
Write a number sentence.

11 + 2 + 6 = 18

Student Model

NAME ______________ DATE ______________

SCORE ______________

Performance Task

Favorite Sport

Jude asked the students in his class their favorite sport. The table shows how many students chose each sport.

Favorite Sport	
Soccer	11
Football	2
Baseball	6

Write your answers on another piece of paper. Show all your work to receive full credit.

Part A

1.MD.4

Write the tallies for each sport.

Favorite Sports		
Sport	Tally	Total
Soccer		11
Football		2
Baseball		6

Performance Task *(continued)*

Part B

1.MD.4

What sport do people like the least?

Answer: baseball

Part C

1.MD.4

Make a bar graph.

Part D

1.MD.4, 1.NBT.4

How many students are in Jude's class?
Write a number sentence.

20

Page 127 • Life on a Farm

Task Scenario	
Students will use analog and digital clocks to tell what time tasks begin and end and will find the height of corn plants in non-standard units.	
CCSS Content Standard(s)	1.MD.2, 1.MD.3
Mathematical Practices	MP1, MP2, MP4, MP5
Depth of Knowledge	DOK2, DOK3, DOK4

Part	Maximum Points	Scoring Rubric
A	2	Full Credit:  Partial Credit (1 point) will be given for showing "8" but not writing the ":00" No credit will be given for an incorrect hour.
B	2	Full Credit:  Partial Credit (1 point) will be given for a correct hour hand but an incorrect minute hand. No credit will be given for an incorrect hour hand.
C	2	Full Credit:  Partial Credit (1 point) will be given for the correct hour (5) but an incorrect or missing minute. No credit will be given for an incorrect hour.

Part	Maximum Points	Scoring Rubric
D	3	Full Credit: Days / Cubes: Monday 5; Tuesday 9; Wednesday 13; Thursday 17 Partial Credit (1 point) will be given for each correct answer (9, 13, 17). No credit will be given for 3 incorrect answers.
TOTAL	9	

Days	Cubes
Monday	5
Tuesday	9
Wednesday	13
Thursday	17

NAME ____________ DATE ____________

SCORE ____________

Performance Task

Life on a Farm

A farmer is growing corn. He needs to measure how tall it grows.

Write your answers on another piece of paper. Show all your work to receive full credit.

Part A I.MD.3

The farmer goes to the field at 6:00. It takes two hours to measure the corn. What time does he finish? Show the time on the clock.

Part B I.MD.3

The farmer finishes eating lunch at 1:00. He spent one hour eating. What time did he start eating lunch? Draw the time on the clock.

Performance Task *(continued)*

Part C I.MD.3

The farmer picks some corn for dinner. He starts picking corn at the time shown on the clock.

He spends a half an hour picking corn. What time does he finish picking corn?

Part D I.MD.2

On Monday the corn was 5 cubes high. It grows 4 cubes higher every day. How many cubes high is it when the farmer measures on Thursday?

Days	Cubes	
Monday	5	+4
Tuesday	9	+4
Wednesday	13	+4
Thursday	17	

Student Model

NAME DATE SCORE

Performance Task

Life on a Farm

A farmer is growing corn. He needs to measure how tall it grows.

Write your answers on another piece of paper. Show all your work to receive full credit.

Part A I.MD.3

The farmer goes to the field at 6:00. It takes two hours to measure the corn. What time does he finish? Show the time on the clock.

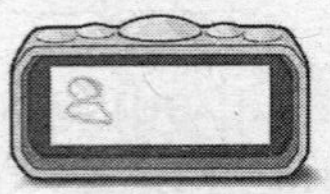

Part B I.MD.3

The farmer finishes eating lunch at 1:00. He spent one hour eating. What time did he start eating lunch? Draw the time on the clock.

Performance Task *(continued)*

Part C I.MD.3

The farmer picks some corn for dinner. He starts picking corn at the time shown on the clock.

He spends a half an hour picking corn. What time does he finish picking corn?

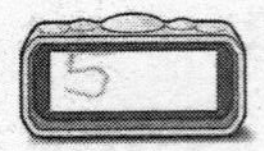

Part D I.MD.2

On Monday the corn was 5 cubes high. It grows 4 cubes higher every day. How many cubes high is it when the farmer measures on Thursday?

Days	Cubes
Monday	5
Tuesday	9
Wednesday	13
Thursday	17

NAME .... DATE ....

SCORE ....

Performance Task

Life on a Farm

A farmer is growing corn. He needs to measure how tall it grows.

Write your answers on another piece of paper. Show all your work to receive full credit.

Part A

I.MD.3

The farmer goes to the field at 6:00. It takes two hours to measure the corn. What time does he finish? Show the time on the clock.

Part B

I.MD.3

The farmer finishes eating lunch at 1:00. He spent one hour eating. What time did he start eating lunch? Draw the time on the clock.

Performance Task *(continued)*

Part C

I.MD.3

The farmer picks some corn for dinner. He starts picking corn at the time shown on the clock.

He spends a half an hour picking corn. What time does he finish picking corn?

Part D

I.MD.2

On Monday the corn was 5 cubes high. It grows 4 cubes higher every day. How many cubes high is it when the farmer measures on Thursday?

Days	Cubes
Monday	5
Tuesday	9
Wednesday	13
Thursday	16

Copyright © McGraw-Hill Education. Permission is granted to reproduce for classroom use.

Student Model

NAME ________________ DATE ________________

SCORE ________________

Performance Task

Life on a Farm

A farmer is growing corn. He needs to measure how tall it grows.

Write your answers on another piece of paper. Show all your work to receive full credit.

Part A

I.MD.3

The farmer goes to the field at 6:00. It takes two hours to measure the corn. What time does he finish? Show the time on the clock.

Part B

I.MD.3

The farmer finishes eating lunch at 1:00. He spent one hour eating. What time did he start eating lunch? Draw the time on the clock.

Performance Task *(continued)*

Part C

I.MD.3

The farmer picks some corn for dinner. He starts picking corn at the time shown on the clock.

He spends a half an hour picking corn. What time does he finish picking corn?

Part D

I.MD.2

On Monday the corn was 5 cubes high. It grows 4 cubes higher every day. How many cubes high is it when the farmer measures on Thursday?

Days	Cubes
Monday	5
Tuesday	9
Wednesday	17
Thursday	25

Page 129 • Building a Playground

<table>
<tr><td colspan="3">Task Scenario
Students will use triangles, rectangles, and trapezoids to show various areas of a playground.</td></tr>
<tr><td colspan="2">CCSS Content Standard(s)</td><td>1.G.1, 1.G.2, 1.G.3</td></tr>
<tr><td colspan="2">Mathematical Practices</td><td>MP1, MP2, MP4, MP7</td></tr>
<tr><td colspan="2">Depth of Knowledge</td><td>DOK2, DOK3, DOK4</td></tr>
<tr><th>Part</th><th>Maximum Points</th><th>Scoring Rubric</th></tr>
<tr><td>A</td><td>3</td><td>Full Credit:
7 edges, 7 vertices
Sample answer:
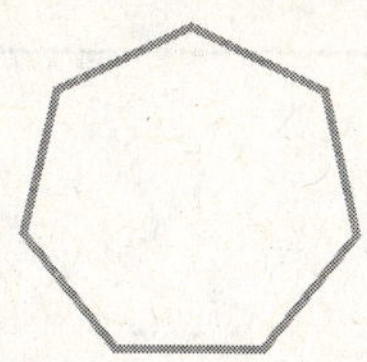
Partial Credit (1 point) will be given for the correct drawing. An additional point will be given for the correct edges OR the correct vertices.
No credit will be given for an incorrect drawing, incorrect edges, and incorrect vertices.</td></tr>
<tr><td>B</td><td>1</td><td>Full Credit:
trapezoid
No credit will be given for an incorrect answer.</td></tr>
<tr><td>C</td><td>2</td><td>Full Credit:
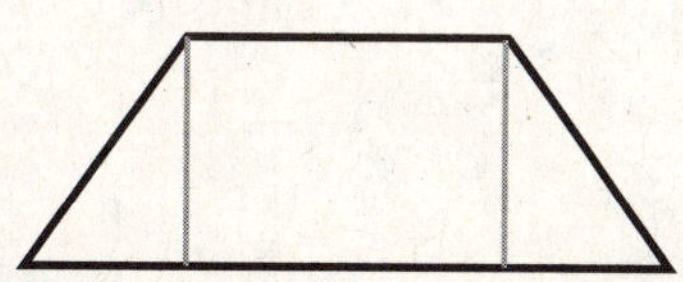
Partial Credit (1 point) will be given for drawing one line but not the other.
No credit will be given for an incorrect division.</td></tr>
</table>

Performance Task Rubrics

Part	Maximum Points	Scoring Rubric
D	2	Full Credit: Sample answer: 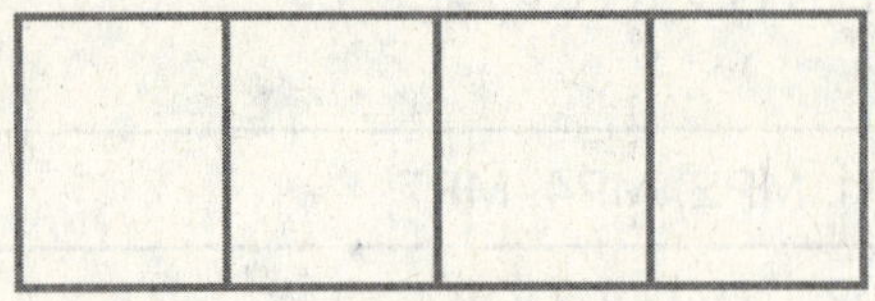 Partial Credit (1 point) will be given for drawing a rectangle but not drawing a correct division **OR** for drawing a different shape but correctly dividing it into fourths. No credit will be given for an incorrect shape AND an incorrect division.
TOTAL	8	

NAME ________________ DATE ________

SCORE ________

Performance Task

Building a Playground

A school wants to build a playground. The shape will be a heptagon. Research what a heptagon looks like.

Write your answers on another piece of paper. Show all your work to receive full credit.

Part A I.G.1

Draw an example of a heptagon. How many edges does it have? How many vertices does it have?

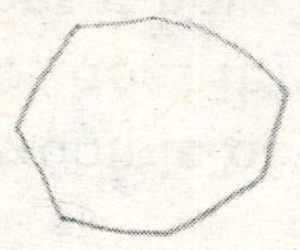

7 edges

7 vertices

Performance Task *(continued)*

Part B I.G.1

The area with slides will be in this shape.

What is this shape called? trapezoid

Part C I.G.1, I.G.2

Grass comes in triangles and rectangles. Draw two lines to make a rectangle and two triangles.

Part D I.G.1, I.G.3

The swing area is a rectangle. It needs divided into fourths. Draw the swing area. Show it divided into fourths.

Student Model

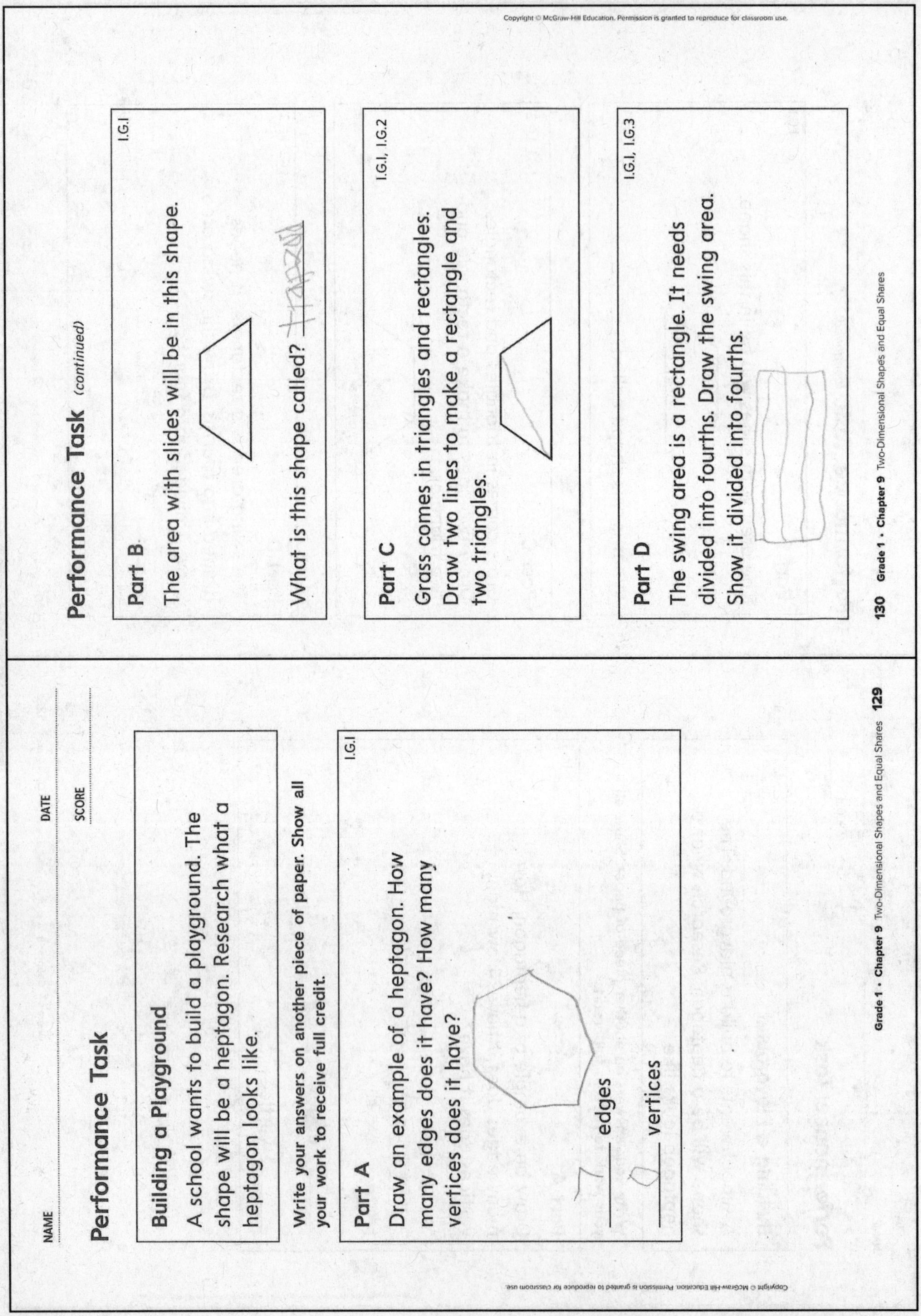

NAME DATE SCORE

Performance Task

Building a Playground

A school wants to build a playground. The shape will be a heptagon. Research what a heptagon looks like.

Write your answers on another piece of paper. Show all your work to receive full credit.

Part A I.G.1

Draw an example of a heptagon. How many edges does it have? How many vertices does it have?

7 edges

8 vertices

Grade 1 • Chapter 9 Two-Dimensional Shapes and Equal Shares 129

Copyright © McGraw-Hill Education. Permission is granted to reproduce for classroom use.

Performance Task *(continued)*

Part B I.G.1

The area with slides will be in this shape.

What is this shape called? trapzoid

Part C I.G.1, I.G.2

Grass comes in triangles and rectangles. Draw two lines to make a rectangle and two triangles.

Part D I.G.1, I.G.3

The swing area is a rectangle. It needs divided into fourths. Draw the swing area. Show it divided into fourths.

130 Grade 1 • Chapter 9 Two-Dimensional Shapes and Equal Shares

Copyright © McGraw-Hill Education. Permission is granted to reproduce for classroom use.

NAME ______________ DATE ______________

SCORE ______________

Performance Task

Building a Playground

A school wants to build a playground. The shape will be a heptagon. Research what a heptagon looks like.

Write your answers on another piece of paper. Show all your work to receive full credit.

Part A I.G.1

Draw an example of a heptagon. How many edges does it have? How many vertices does it have?

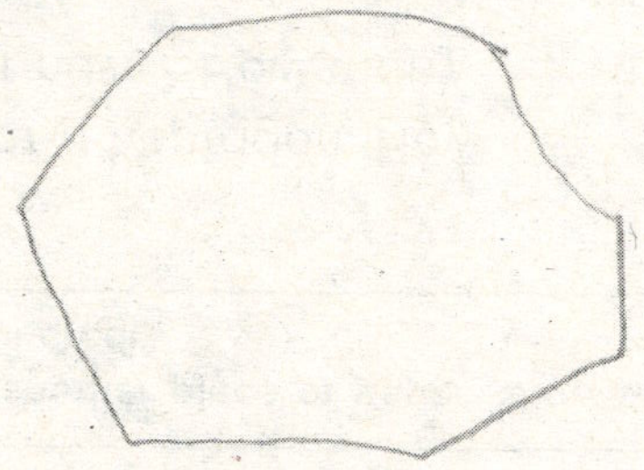

6 edges

6 vertices

Performance Task *(continued)*

Part B I.G.1

The area with slides will be in this shape.

What is this shape called? rectangle

Part C I.G.1, I.G.2

Grass comes in triangles and rectangles. Draw two lines to make a rectangle and two triangles.

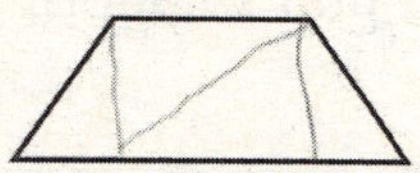

Part D I.G.1, I.G.3

The swing area is a rectangle. It needs divided into fourths. Draw the swing area. Show it divided into fourths.

Student Model

NAME ______ DATE ______

SCORE ______

Performance Task

Building a Playground

A school wants to build a playground. The shape will be a heptagon. Research what a heptagon looks like.

Write your answers on another piece of paper. Show all your work to receive full credit.

Part A

I.G.1

Draw an example of a heptagon. How many edges does it have? How many vertices does it have?

6 edges

5 vertices

Performance Task *(continued)*

Part B

I.G.1

The area with slides will be in this shape.

What is this shape called? ______

Part C

I.G.1, I.G.2

Grass comes in triangles and rectangles. Draw two lines to make a rectangle and two triangles.

Part D

I.G.1, I.G.3

The swing area is a rectangle. It needs divided into fourths. Draw the swing area. Show it divided into fourths.

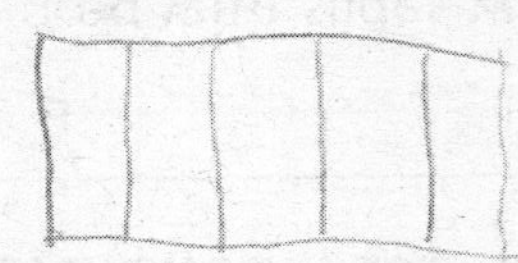

Page 131 • Making a Castle Wall

Task Scenario		
Students will use three-dimensional shapes to build a castle wall and complete patterns.		
CCSS Content Standard(s)		1.G.1, 1.G.2
Mathematical Practices		MP1, MP2, MP4, MP7
Depth of Knowledge		DOK2, DOK3
Part	**Maximum Points**	**Scoring Rubric**
A	3	Full Credit: cylinder, rectangular prism, and cone Partial Credit (2 points) will be given for having three of the four marked correctly. 1 point will be given for having two of the four marked correctly. No credit will be given for 0 or 1 correct markings.
B	3	Full Credit: 0, 1, 8 cylinder Partial Credit (2 points) will be given for having three of four correct answers. 1 point will be given for having two of four correct answers. No credit will be given for 0 or 1 correct answers.
C	2	Full Credit: rectangular prism No credit will be given for an incorrect answer.
TOTAL	8	

NAME ______________________ DATE ______________

SCORE ______________

Performance Task

Making a Castle Wall

Jerome is building a castle wall out of blocks.

Write your answers on another piece of paper. Show all your work to receive full credit.

Part A

I.G.2

The first part of the wall looks like this. Circle the names of the shapes that are used to make the wall.

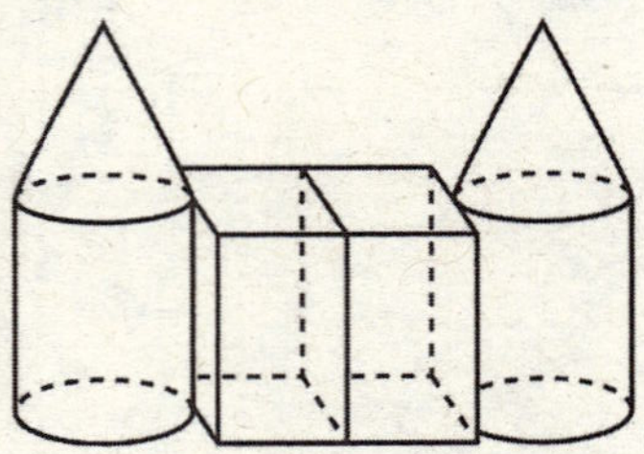

cylinder cube

rectangular prism cone

Performance Task *(continued)*

Part B

I.G.1

How many vertices does each shape have?

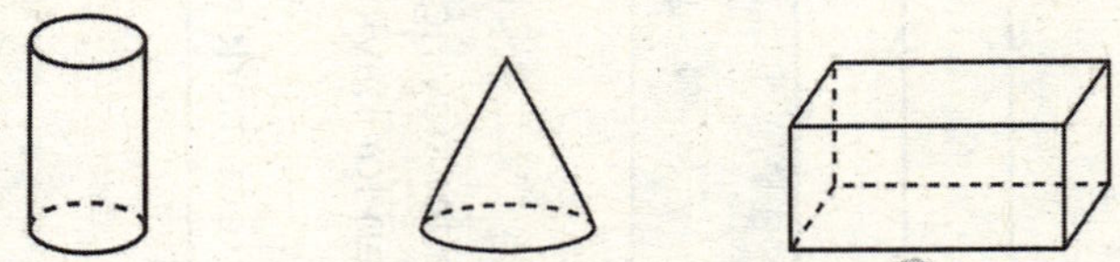

0 ______ 1 ______ 8 ______

Which one has the least? cylinder ______

Part C

I.G.1

The wall will keep going with the same pattern. Circle the shape that comes next.

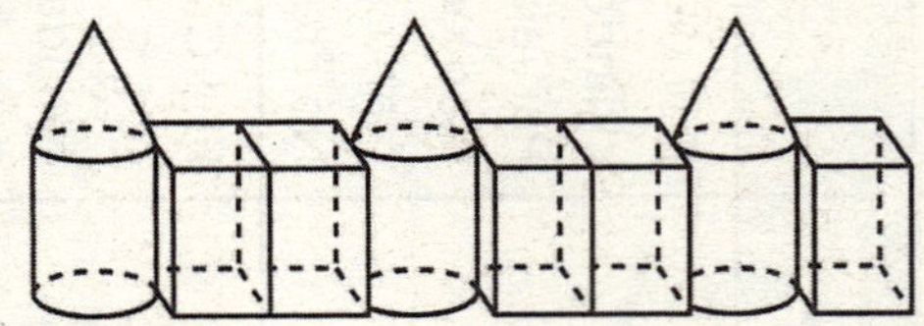

______

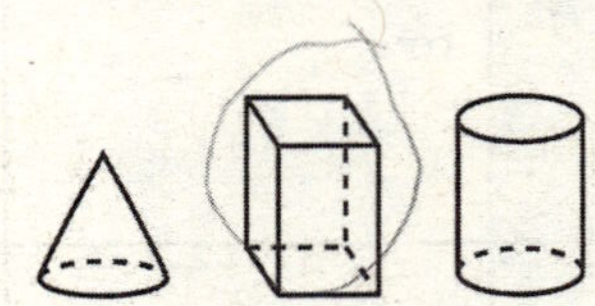

NAME ______________________ DATE ____________

SCORE ____________

Performance Task

Making a Castle Wall

Jerome is building a castle wall out of blocks.

Write your answers on another piece of paper. Show all your work to receive full credit.

Part A I.G.2

The first part of the wall looks like this. Circle the names of the shapes that are used to make the wall.

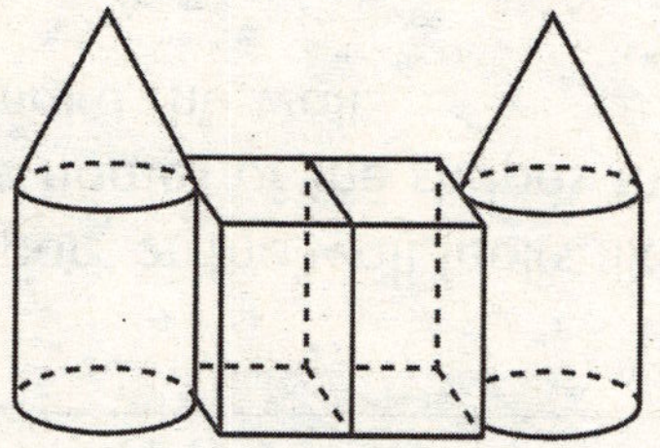

cylinder (circled) cube

rectangular prism (circled) cone (circled)

Performance Task *(continued)*

Part B I.G.1

How many vertices does each shape have?

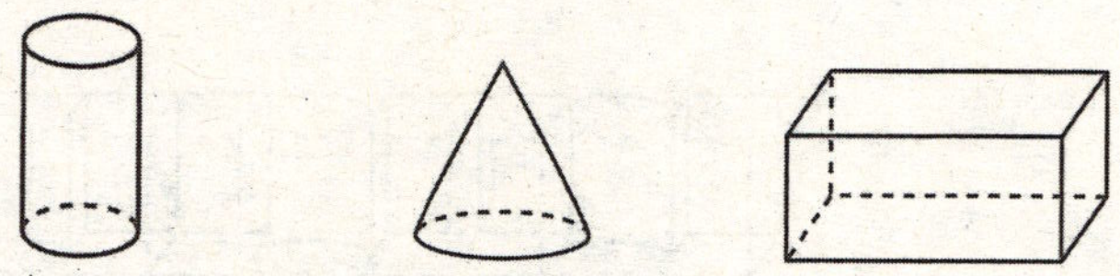

2 ______ 1 ______ 6 ______

Which one has the least? cone ______

Part C I.G.1

The wall will keep going with the same pattern. Circle the shape that comes next.

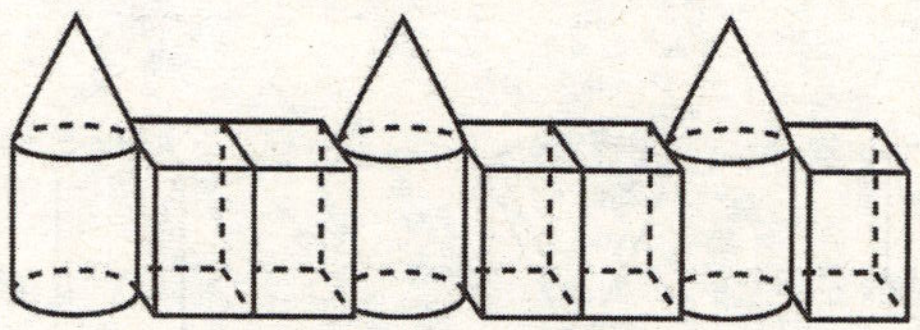

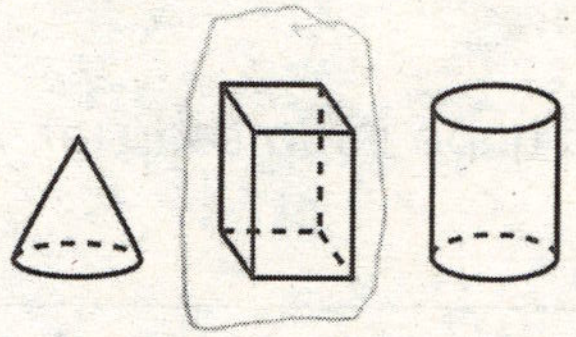

Student Model

NAME ____________ DATE ____________

SCORE ____________

Performance Task

Making a Castle Wall

Jerome is building a castle wall out of blocks.

Write your answers on another piece of paper. Show all your work to receive full credit.

Part A

I.G.2

The first part of the wall looks like this. Circle the names of the shapes that are used to make the wall.

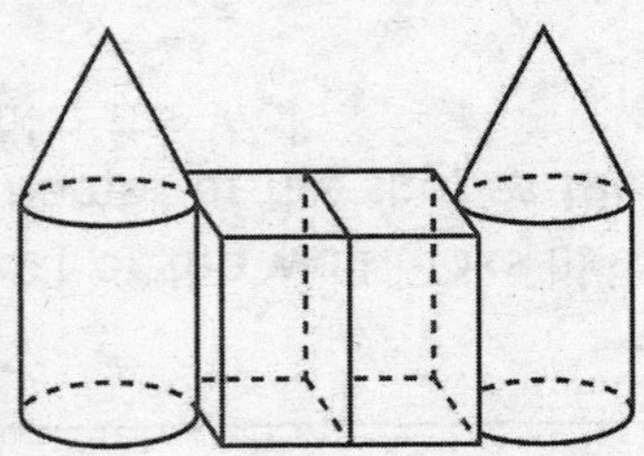

cylinder

cube

rectangular prism

cone

Performance Task *(continued)*

Part B

I.G.1

How many vertices does each shape have?

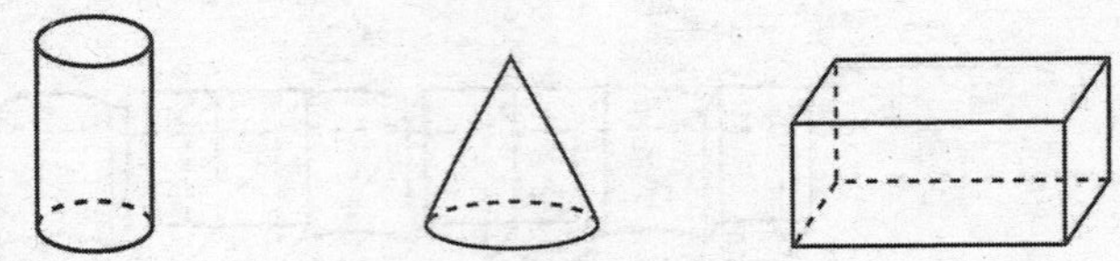

1 ______ 1 ______ 6 ______

Which one has the least? cylinder + cone

Part C

I.G.1

The wall will keep going with the same pattern. Circle the shape that comes next.

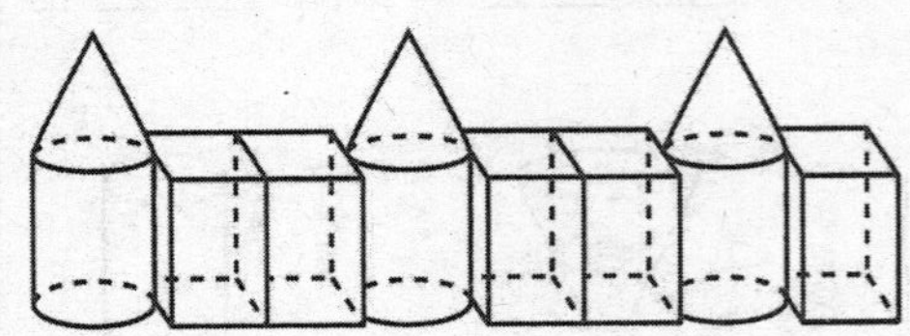

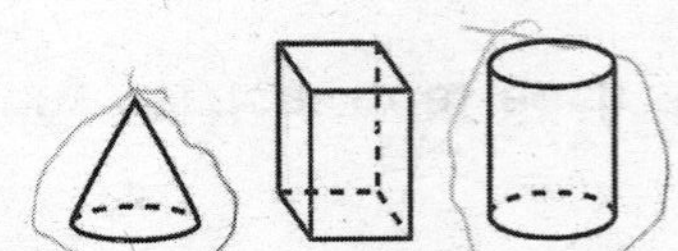

NAME ________ DATE ________

SCORE ________

Performance Task

Making a Castle Wall

Jerome is building a castle wall out of blocks.

Write your answers on another piece of paper. Show all your work to receive full credit.

Part A

I.G.2

The first part of the wall looks like this. Circle the names of the shapes that are used to make the wall.

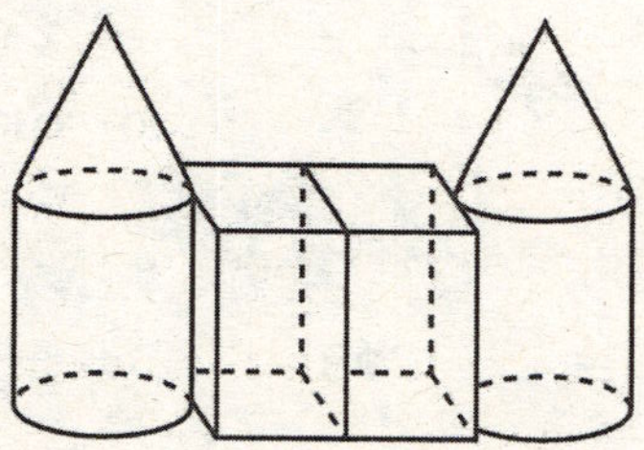

cylinder cube

rectangular prism cone

Performance Task *(continued)*

Part B

I.G.1

How many vertices does each shape have?

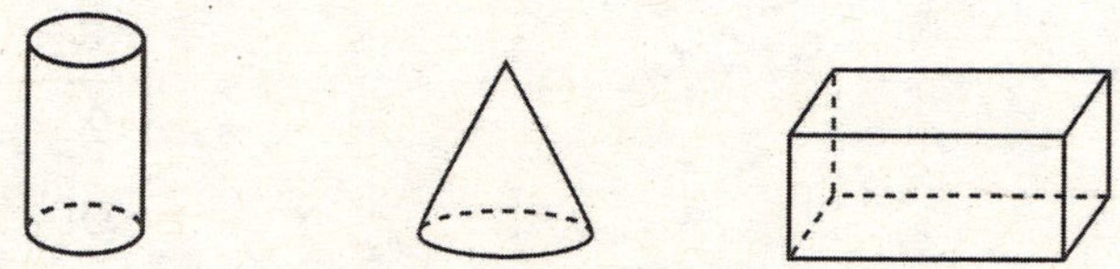

3 ____ 2 ____ 8 ____

Which one has the least? the cone ____

Part C

I.G.1

The wall will keep going with the same pattern. Circle the shape that comes next.

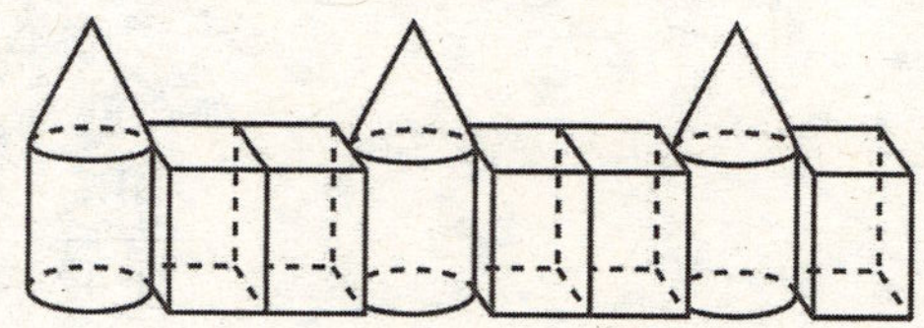

____

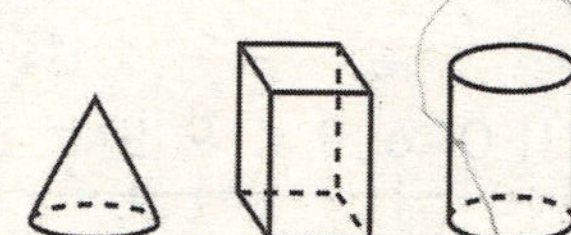

Student Model

NAME ________ DATE ________

SCORE ________

Benchmark Test I

1. Maddie has 4 pennies in her pocket. She has 8 pennies on her desk. How many pennies does Maddie have in all? I.OA.6

4¢ + 8¢ = 12 ¢

2. Add zero. I.OA.3

7 + 0 = 7

3. True or false? I.OA.I, I.OA.7

Tyler has 8 cars. He gives I to his brother. Now Tyler has 7 cars.

(True) False

4. Use the number line to solve. Write the sum. I.OA.6

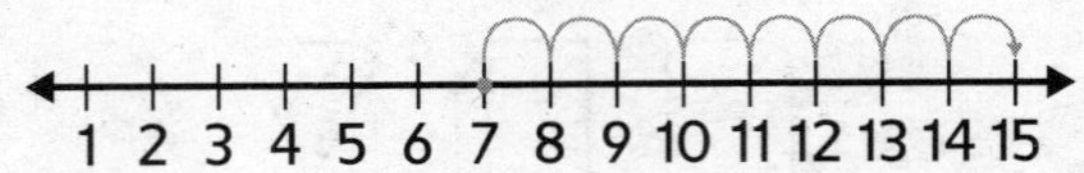

$$\begin{array}{r} 7 \\ +\ 8 \\ \hline 15 \end{array}$$

5. True or false?

Zachary has 7 blocks. He picks up I more block. Zachary has 8 blocks in all. I.OA.I, I.OA.7

(True) False

6. Subtract. I.OA.6

$$\begin{array}{r} 10 \\ -\ 6 \\ \hline 4 \end{array}$$

7. Find the matching addition number sentence. I.OA.I

A. 3 + 4 = 6

(B. 3 + 4 = 7) — circled

C. 2 + 4 = 6

8. Circle the correct doubles facts. I.OA.6

6 + 6 = 11 (8 + 8 = 16) — circled

4 + 4 = 6 (9 + 9 = 18) — circled

9. Write the related subtraction fact. I.OA.6

7 − 6 = 1

_7_ − _1_ = _6_

10. Abigail's teacher asked her to add 7 + 5. She makes 10 to add. Which number sentence does Abigail use? I.OA.6

10 + 1 = 11

(10 + 2 = 12) — circled

10 + 3 = 13

11. Find the missing part. I.OA.I

Part	Part
2	5
Whole	
7	

12. Write a number sentence. I.OA.I

There were 6 pieces of pizza. Andre eats some of the pieces. There are 3 left. How many did Andre eat?

_6_ (−) _3_ = _3_ pieces

13. True or false? Mr. Valencia has 8 red ties and 6 blue ties. He has 8 blue socks and 6 black socks. Mr. Valencia has the same number of ties as socks. I.OA.3, I.OA.7

(True) False

14. Geraldine has 8 pencils. Circle the ways that will make 8 pencils. I.OA.6

(2 + 6) 4 + 3

5 + 2 (0 + 8)

15. Mrs. Clark buys 5 red balloons, 7 orange balloons, and 3 yellow balloons. How many balloons did she buy in all? I.OA.2

__5__ + __7__ + __3__ = __15__ balloons

16. Find the matching subtraction number sentence. I.OA.1, I.OA.5

(A. 8 − 5 = 3)

B. 7 − 5 = 3

C. 5 − 8 = 3

17. Antwan read 7 books in January. He read 9 books in February. How many books did he read in all? I.OA.6

__16__ books

18. Circle two numbers that will make 10. I.OA.6

2 6 (3)

(7) 1 5

Copyright © McGraw-Hill Education. Permission is granted to reproduce for classroom use.

19. Add. 1.OA.6

$$\begin{array}{r} 7 \\ +\,7 \\ \hline 14 \end{array}$$

20. Find the related subtraction sentence. 1.OA.6

$$10 - 4 = 6$$

A. $6 - 4 = 2$

B. $10 - 2 = 8$

C. $10 - 6 = 4$

Page 140 • Collecting Cans

Task Scenario		
Students will use addition and subtraction to determine the number of cans Paul has to recycle.		
CCSS Content Standard(s)		1.OA.1, 1.OA.6
Mathematical Practices		MP1, MP2, MP4
Depth of Knowledge		DOK2, DOK3
Part	**Maximum Points**	**Scoring Rubric**
A	3	Full Credit: $6 + 3 = 9$ cans Partial Credit (1 point) will be given for a correct answer without a correct number sentence. No credit will be given for an incorrect answer.
B	3	Full Credit: $7 - 4 = 3$ cans Partial Credit (1 point) will be given for a correct answer without a correct number sentence. No credit will be given for an incorrect answer.
C	2	Full Credit: $7 - 3 = 4$ cans Partial Credit (1 point) will be given for a correct answer without a correct number sentence. No credit will be given for an incorrect answer.
TOTAL	8	

NAME ________________ DATE ________

SCORE ________

Benchmark Test 2

1. Circle the correct answer. I.NBT.4

$$\begin{array}{r} 30 \\ +\ 30 \\ \hline \end{array}$$

A. 6

B. 16

C. 60 (circled)

2. Carlina has the boxes of tissues shown. Circle a group of ten boxes. Write how many more. How many boxes does Carlina have in all? I.NBT.2b

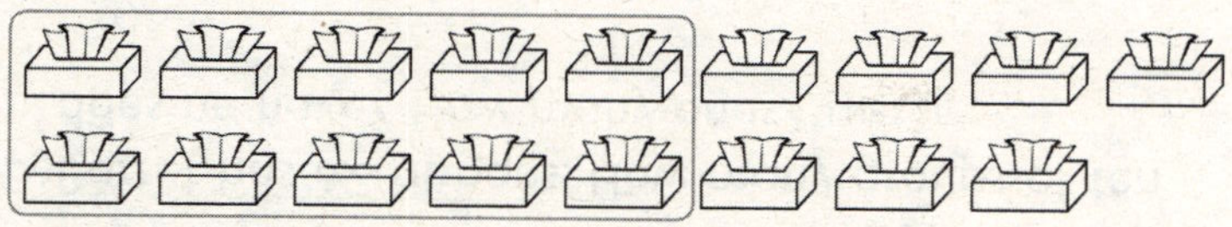

10 and 7 more 17 boxes

3. True or false? 4 tens and 7 ones is the same as forty-seven. I.NBT.5

True (circled) False

4. 20 + ____ = 90. Mark the related subtraction fact. I.NBT.6

- ☐ 20 + 60 = 90
- ■ 90 − 20 = 70
- ☐ 90 − 20 = 60
- ☐ 70 + 20 = 90

5. Ted starts counting backwards from 17. He says, "17, 16, 15, 14, 13, 12." Circle the number sentence is he trying to solve. I.OA.5

17 − 5 = 12 (circled) 17 − 4 = 12

17 − 4 = 13 17 − 3 = 13

6. Are the number sentences part of the fact family? Circle yes or no. I.NBT.6

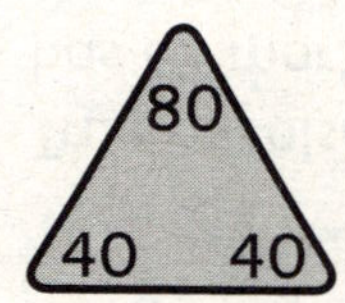

yes (circled) no 40 + 40 = 80

yes (circled) no 80 − 40 = 40

yes no (circled) 40 − 80 = 40

Benchmark Tests

7. Damian has 53 stamps. He buys 30 more stamps. How many stamps does Damian have in all? I.NBT.4

$$\begin{array}{r} 53 \\ +\ 30 \\ \hline 83 \end{array}$$ stamps

8. Barret has 62 ribbons. How many groups of ten does he have? How many ones? I.NBT.2a

__6__ tens and __2__ ones

9. There are 16 blocks. Katherine loses 5 of them. How many blocks does she have left? Use the number line to help you subtract. I.OA.5

__11__ blocks

10. Use the number line to subtract. Write the difference. I.NBT.6

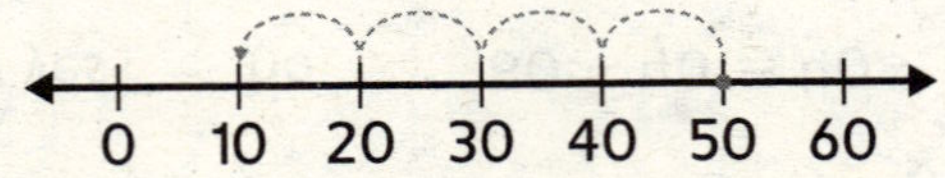

50 – 40 = __10__

11. True or false? Deangelo has 43 marbles. Kumar has 63 marbles. Kumar has 20 more marbles than Deangelo. I.NBT.5

(True) False

12. Elise is thinking of a number between 99 and 101. What is her number? I.NBT.1

A. 98

(B. 100)

C. 102

13. Manuel saw two kinds of animals at the zoo. He saw 14 animals in all. Circle the kinds of animals Manuel saw. 1.OA.3, 1.OA.7

6 Leopards   8 Lions   4 Tigers

14. Mr. Huan drinks 10 cups of water every day. How many cups does he drink in 5 days? Fill in the table. 1.NBT.2c

Days	Cups
1	10
2	20
3	30
4	40
5	50

15. Miranda buys a greeting card that costs 55¢. She only used nickels to pay. How many nickels did Miranda use? 1.NBT.1

__11__ nickels

16. Luciana has 13 granola bars. She gives away 5 of them. How many granola bars does she still have? Write an addition sentences and a subtraction sentence. 1.OA.6

__5__ + __8__ = __13__

__13__ − __5__ = __8__ granola bars

17. Gianna is thinking of a number. When she adds 3 to the number, she gets 10. What is her number? 1.OA.4

3 + __7__ = 10

18. Mrs. Klein has 16 pencils. She buys 8 more. How many pencils does she have in all? Circle the ones to show regrouping. Write your answer. 1.NBT.4

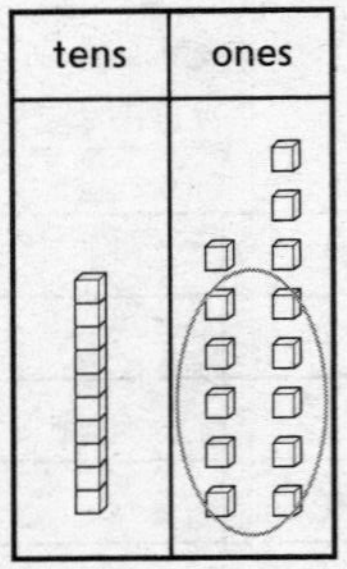

24 pencils

19. Which subtraction problem has the same answer as 16 – 7? 1.OA.6

(10 – 1)

10 – 2

10 – 3

20. Mia has 4 red cups and some orange cups. She has 14 cups in all. How many orange cups does Mia have? 1.OA.4

10 orange cups

Copyright © McGraw-Hill Education. Permission is granted to reproduce for classroom use.

Page 148 • United States Senators

Task Scenario Students will use two-digit addition and subtraction to answer questions about United States senators.	
CCSS Content Standard(s)	1.NBT.4, 1.NBT.6
Mathematical Practices	MP1, MP2, MP4
Depth of Knowledge	DOK2, DOK3

Part	Maximum Points	Scoring Rubric
A	4	Full Credit: 70 senators 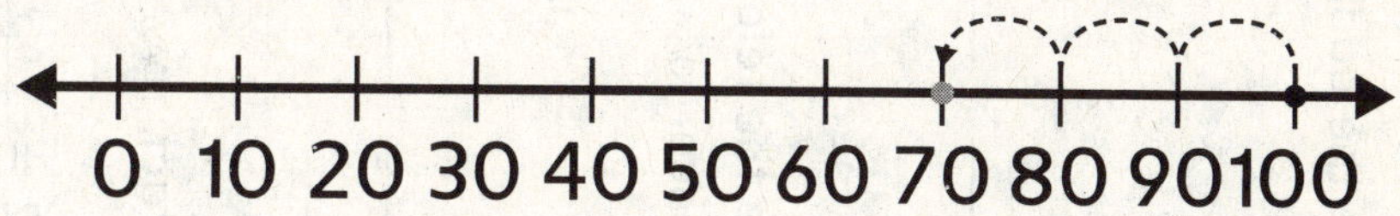 Partial Credit (1 point) will be given for a correct answer without a correct number sentence. Another point will be given for having the correct two dots on the number line without the appropriate arches. No credit will be given for an incorrect answer.
B	2	Full Credit: 25 + 3 = 28 senators Partial Credit (1 point) will be given for a correct answer without a correct number sentence. No credit will be given for an incorrect answer.
C	2	Full Credit: 16 + 20 = 36 senators Partial Credit (1 point) will be given for a correct answer without a correct number sentence. No credit will be given for an incorrect answer.
TOTAL	8	

NAME ____________ DATE ____________

SCORE ____________

Benchmark Test 3

1. Olivia grew a green bean plant. It had 9 beans on it. Olivia ate 5 green beans. How many green beans are left? I.OA.1

A. 9 − 5 = 3 green beans

B. 9 − 5 = 4 green beans

C. 9 − 4 = 5 green beans

2. Find the missing part of 10. I.OA.4, I.OA.8

Part	Part
6	4
Whole	
10	

3. Write the addition number sentence. I.OA.1, I.OA.1

3 + 2 = 5

4. Write the related subtraction number sentence. I.OA.6

4 − 3 = 1

4 − 1 = 3

Subtract. I.OA.6

5. 6 − 2 = 4

6. 10 − 3 = 7

7. True or false? 1.OA.1, 1.OA.7

Jamal has 8 apples. He eats 3. Now Jamal has 5 apples.

(True) False

8. Make 10 to add. 1.OA.3, 1.OA.6

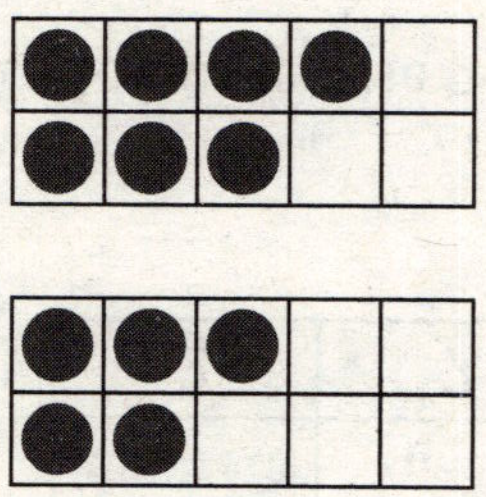

$$\begin{array}{r} 7 \\ +\,5 \\ \hline \end{array} \longrightarrow \begin{array}{r} 10 \\ +\,2 \\ \hline 12 \end{array}$$

9. Circle the greater number. Count on to add. 1.OA.5, 1.OA.6

3 + (9) = 12

Find the sum. 1.OA.6

10. $\begin{array}{r} 8 \\ +\,5 \\ \hline 13 \end{array}$

11. $\begin{array}{r} 6 \\ +\,9 \\ \hline 15 \end{array}$

12. 14 − 6 1.OA.6

4 2

14 − 4 = 10

10 − 2 = 8

So, 14 − 6 = 8

13. 18 − 9 = 9 1.OA.6

Benchmark Tests

14. Complete the fact family. 1.OA.3, 1.OA.6

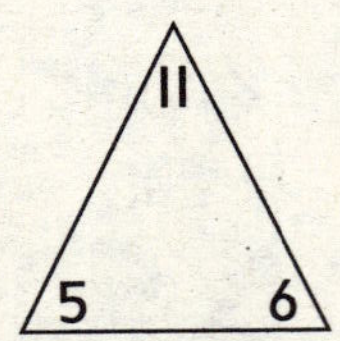

5 + 6 = 11   11 − 6 = 5

6 + 5 = 11   11 − 5 = 6

15. Complete the place value chart. 1.NBT.2

tens   hundreds   ones

hundreds	tens	ones
2	7	8

16. Compare. Use >, <, or =. 1.NBT.3

34 (>) 28

17. Josiah bakes 4 groups of ten rolls. He bakes 3 more rolls. How many rolls did he bake in all? 1.NBT.4

43

18. Read the numbers. What number is missing? 1.NBT.1

117, ______, 119, 120

A. 121

B. 115

C. 118 (circled)

19. Subtract. 1.NBT.2c, 1.NBT.6

7 tens − 5 tens = 2 tens

70 − 50 = 20

20. Add. I.NBT.4

	tens	ones
	2	2
+	0	4
	2	6

21. Count on to add. I.NBT.4

86 + 3 = 89

22. Complete the tally chart. I.MD.4

Favorite Pet		
Type	Tally	Total
Cat	卌 |||	8
Dog	卌 |	6
Fish	||||	4

23. Use the tally chart in Exercise 22 to make a bar graph. I.MD.4

24. Use the graphs above to answer the question. How many people chose fish and cats? I.MD.4

12 people

25. Order the objects by length. Write 1 for long, 2 for longer, and 3 for longest. I.MD.1

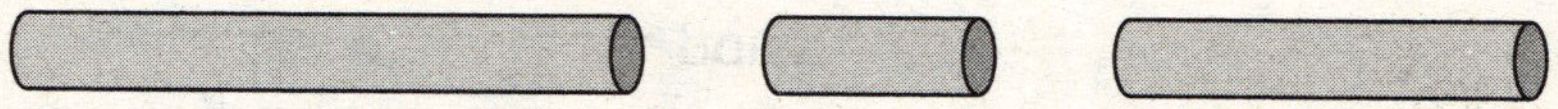

3   1   2

Benchmark Tests

26. Write the time on the digital clock. I.MD.3

half past 4

27. Write the time on the clock. I.MD.3

Piper gets up at 7:30. Her bus comes 1 hour later. What time does the bus come?

28. Write how many sides and vertices. I.G.1

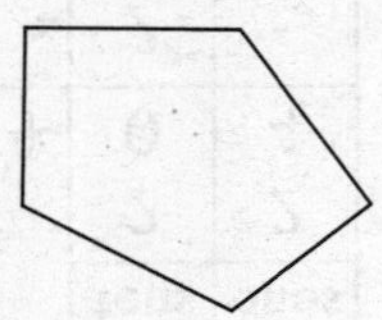

5 sides

5 vertices

29. Write how many parts are shaded. I.G.3

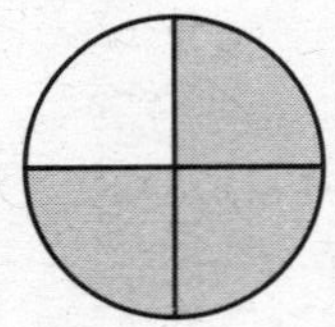

3 of 4 parts

30. Circle the shape that is not used to build the composite shape below. I.G.2

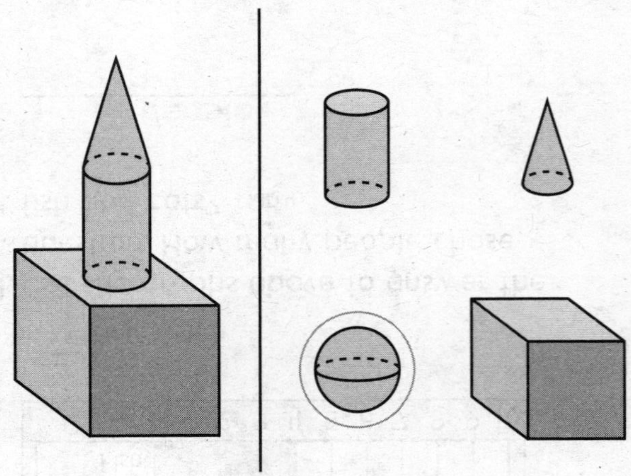

31. Circle the name of the shape. Write the number of faces and vertices. I.G.1

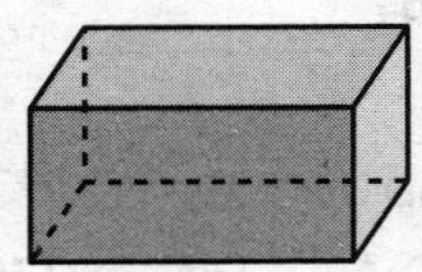

cone

rectangular prism

6 faces

8 vertices

Page 159 • Building Blocks

Task Scenario Students will describe and compare shapes, solve problems by finding a pattern, and recognize how a composite shape is created.		
CCSS Content Standard(s)	1.G.1, 1.G.2	
Mathematical Practices	MP1, MP2, MP6, MP7, MP8	
Depth of Knowledge	DOK1, DOK2, DOK3	

Part	Maximum Points	Scoring Rubric
A	3	Full Credit: The student correctly identifies 6 faces in a cube and 2 in a cylinder and correctly identifies that a cube has more faces. Partial Credit (2 points) will be given if the student correctly identifies 6 faces in a cube and 2 in a cylinder but identifies that a cylinder has more faces **OR** correctly identifies 6 faces in a cube and correctly identifies that a cube has more faces, but incorrectly identifies the number of faces in a cylinder **OR** correctly identifies 2 faces in a cylinder and correctly identifies that a cube has more faces, but incorrectly identifies the number of faces in a cube. Partial Credit (1 point) will be given if the student correctly identifies 6 faces in a cube but incorrectly identifies the number of faces in a cylinder and identifies that a cylinder has more faces **OR** correctly identifies 2 faces in a cylinder but incorrectly identifies the number of faces in a cube and identifies that a cylinder has more faces **OR** correctly identifies that a cube has more faces, but incorrectly identifies the number of faces in a cube and in a cylinder. No credit will be given for an incorrect answer.

Performance Task Rubric

Part	Maximum Points	Scoring Rubric
B	3	The student correctly identifies 8 vertices in a cube and 1 vertex in a cone and correctly identifies that a cube has more faces. Partial Credit (2 points) will be given if the student correctly identifies 8 vertices in a cube and 1 in a cone but identifies that a cylinder has more vertices **OR** correctly identifies 8 vertices in a cube and correctly identifies that a cube has more vertices, but incorrectly identifies the number of vertices in a cone **OR** correctly identifies 1 vertex in a cone and correctly identifies that a cube has more vertices, but incorrectly identifies the number of vertices in a cone. Partial Credit (1 point) will be given if the student correctly identifies the number of vertices in a cube but incorrectly identifies the number of vertices in a cone and identifies that a cone has more vertices **OR** correctly identifies the number of vertices in a cone but incorrectly identifies the number of vertices in a cube and identifies that a cone has more vertices **OR** correctly identifies that a cube has more vertices, but incorrectly identifies the number of vertices in a cube and in a cone. No credit will be given for an incorrect answer.
C	1	Full Credit: The student correctly identifies the missing solid shapes. No credit will be given for an incorrect answer.
D	4	Full Credit: The student correctly matches all four solid shapes to their place on the composite figure. Partial Credit (3 points) will be given if the student correctly matches three of the shapes. Partial Credit (2 points) will be given for correctly matching two of the solid shapes. Partial Credit (1 point) will be given for correctly matching two of the solid shapes. No credit will be given for an incorrect answer.
TOTAL	11	

NAME ______________ DATE ______________

SCORE ______________

Benchmark Test 4

1. Write the addition number sentence. I.OA.I

3 + 3 = 6

2. 12 − 5 = 7 I.OA.6

3. Write the time on the digital clock. I.MD.3

half past two

4. 17 − 9 I.OA.6

7 | 2

17 − 7 = 10

10 − 2 = 8

So, 17 − 9 = 8.

5. Count on to add. I.NBT.4

72 + 5 = 77

6. Janice had a pizza with 8 pieces in it. Janice ate 3 pieces. How many pieces are left? I.OA.I

A. 8 − 3 = 5 pieces (circled)

B. 8 − 5 = 4 pieces

C. 8 − 5 = 3 pieces

7. Complete the place value chart. I.NBT.2

tens	hundreds	ones

hundreds	tens	ones
1	6	4

8. Write the related subtraction number sentence. I.OA.6

6 − 2 = 4

6 − 4 = 2

9. Write the time on the clock. I.MD.3

Veronica starts reading at 6:30. She goes to bed one hour later. What time does she go to bed?

10. Write how many sides and vertices. I.G.1

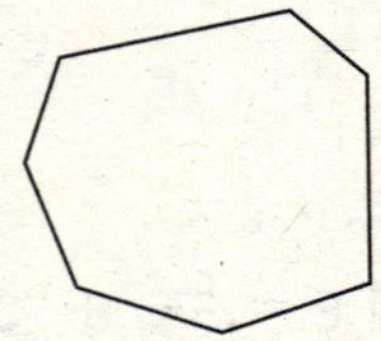

7 sides

7 vertices

11. Complete the fact family. I.OA.3, I.OA.6

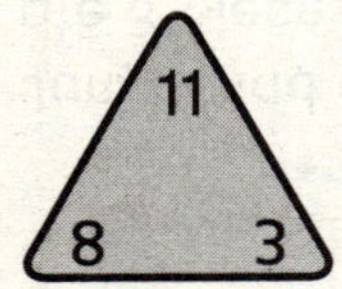

8 + 3 = 11

3 + 8 = 11

11 − 8 = 3

11 − 3 = 8

12. Find the missing part of 10. I.OA.4, I.OA.8

Part	Part
7	3
Whole	
10	

13. Complete the tally chart. I.MD.4

Favorite Season		
Type	**Tally**	**Total**
Summer	𝍸 ||||	9
Fall	𝍸	5
Winter	𝍸 |	6
Spring	||	2

14. Use the tally chart in Exercise 13 to make a bar graph. I.MD.4

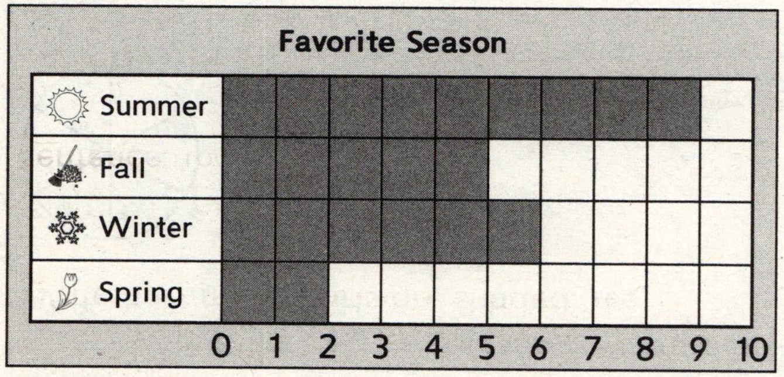

15. Use the graphs above to answer the question. How many people chose fall or winter? I.MD.4

11

Subtract. I.OA.6

16. 7 – 5 = 2

17. 10 – 5 = 5

18. Add. I.NBT.4

	tens	ones
	3	1
+		5
	3	6

19. Compare. Use >, <, or =. I.NBT.3

28 (<) 82

20. True or false? 1.OA.1, 1.OA.7

Brayden had 9 books. He took 4 back to the library. Now Brayden has 5 books.

(True) False

21. Write how many parts are shaded. 1.G.3

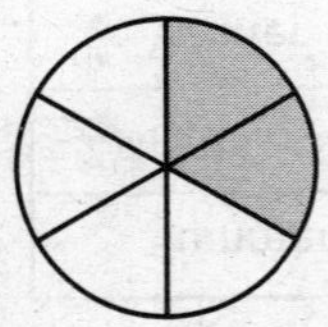

2 of 6 parts

22. Marcus buys 4 groups of ten baseball cards. He buys 2 more cards. How many cards did he buy in all? 1.NBT.4

42 cards

23. Circle the name. Write the number of faces and vertices. 1.G.1

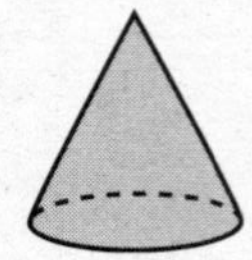

(cone) cube

cylinder rectangular prism

1 faces 1 vertices

24. Make 10 to add. 1.OA.3, 1.OA.6

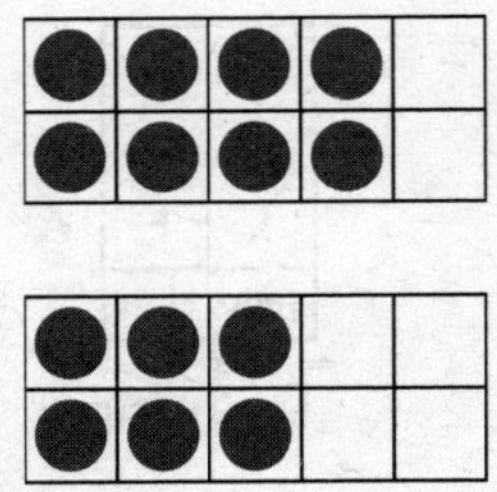

$$\begin{array}{r} 8 \\ +\ 6 \\ \hline 14 \end{array} \longrightarrow \begin{array}{r} 10 \\ +\ 4 \\ \hline 14 \end{array}$$

25. Find the sum. 1.OA.6

$$\begin{array}{r} 7 \\ +\ 4 \\ \hline 11 \end{array}$$

26. Find the sum. I.OA.6

$$\begin{array}{r} 6 \\ +\ 8 \\ \hline 14 \end{array}$$

27. Circle the shape that is not used to build the composite shape below. I.G.2

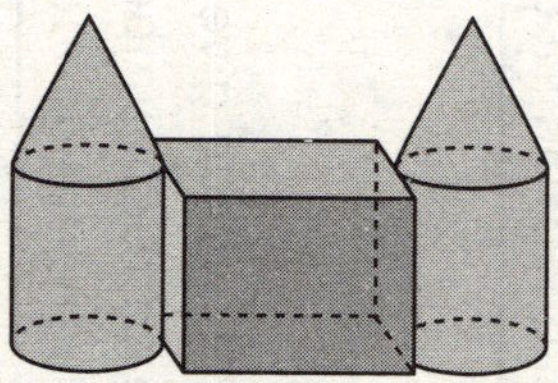

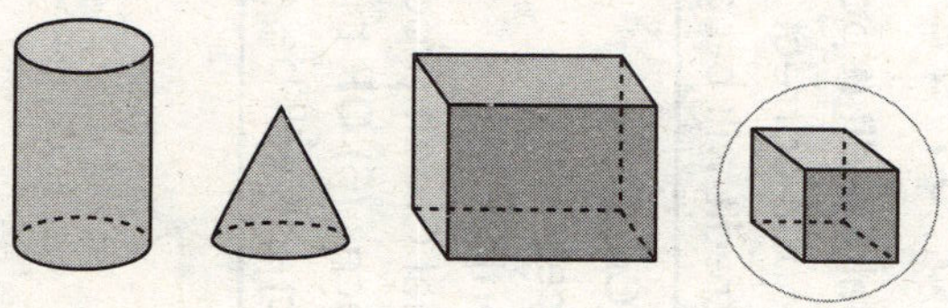

28. Subtract. I.NBT.2c, I.NBT.6

8 tens − 3 tens = 5 tens

80 − 30 = 50

29. Circle the greater number. Count on to add. I.OA.5, I.OA.6

5 + (8) = 13

30. Read the numbers. What number is missing? I.NBT.1

108, 109, ____, 111

A. 100

B. 101

C. 110 (circled)

31. Order the objects by length. Write 1 for shortest, 2 for shorter, and 3 for short. I.MD.1

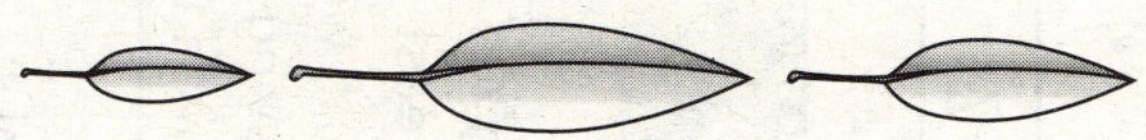

1 3 2

Page 171 • Donating Books for a Library

Task Scenario Students will use addition and subtraction of two digit numbers, number lines, time to the half hour, and three-dimensional shapes to answer questions about library books.	
CCSS Content Standard(s)	1.NBT.3, 1.NBT.4, 1.OA.5, 1.MD.3, 1.G.1
Mathematical Practices	MP1, MP2, MP4, MP7
Depth of Knowledge	DOK2, DOK3, DOK4

Part	Maximum Points	Scoring Rubric
A	2	Full Credit: Deshawn collected the most. Kim collected the least. Partial Credit (1 point) will be given for having the correct answer for the most **OR** having the correct answer for the least. No credit will be given for two incorrect responses.
B	2	Full Credit: 36 + 20 = 56 books Partial Credit (1 point) will be given for a correct answer (56) without a correct number sentence. No credit will be given for an incorrect answer.
C	2	Full Credit: 32 books 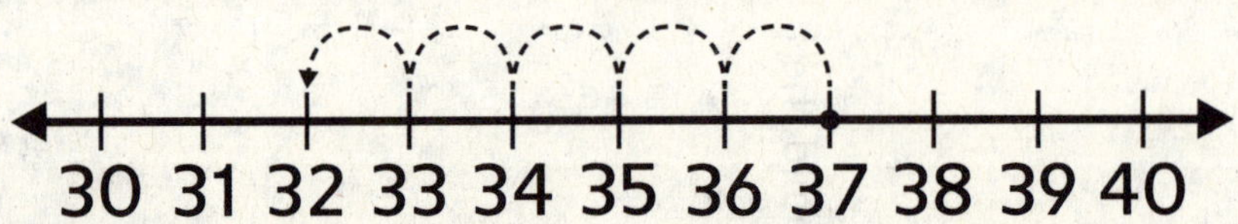 Partial Credit (1 point) will be given for a correct answer (32) without a correct number line. No credit will be given for an incorrect answer.
D	2	Full Credit: 6 faces 8 vertices Partial Credit (1 point) will be given for the correct number of faces (6) **OR** the correct number of vertices (8). No credit will be given for two incorrect answers.

Part	Maximum Points	Scoring Rubric
E	2	Full Credit:  Partial Credit (1 point) will be given for having a correct minute hand **OR** a correct hour hand. No credit will be given for having both hands incorrectly placed.
TOTAL	10	